# Mobile Communications

Edited by
**Fraidoon Mazda**
MPhil DFH CEng FIEE

With specialist contributions

Focal Press
An imprint of Butterworth-Heinemann
Linacre House, Jordan Hill, Oxford OX2 8DP
A division of Reed Educational and Professional Publishing Ltd

*R* A member of the Reed Elsevier plc group

OXFORD BOSTON JOHANNESBURG
NEW DELHI SINGAPORE MELBOURNE

First published 1996

© Reed Educational and Professional Publishing Ltd 1996

All rights reserved. No part of this publication
may be reproduced in any material form (including
photocopying or storing in any medium by electronic
means and whether or not transiently or incidentally
to some other use of this publication) without the
written permission of the copyright holder except in
accordance with the provisions of the Copyright,
Designs and Patents Act 1988 or under the terms of a
licence issued by the Copyright Licensing Agency Ltd,
90 Tottenham Court Road, London, England, W1P 9HE.
Applications for the copyright holder's written permission
to reproduce any part of this publication should be addressed
to the publishers

**British Library Cataloguing in Publication Data**
Mazda, Fraidoon F
   Mobile Communications
   I. Title
   621.382

D
621.382
TEL

**ISBN 0 2405 1458 0**

**Library of Congress Cataloguing in Publication**
Mazda, Fraidoon F.
   Mobile Communications/Fraidoon Mazda
   p. cm.
   Includes bibliographical references and index.
   **ISBN 02405 145x x**
   1. Telecommunications. I. Title
   TK5101.M37 1993
   621.382–dc20
Printed and bound in Great Britain

92-27846
CIP

# Contents

| | |
|---|---|
| Preface | vii |
| List of contributors | ix |
| 1. Radio paging | 1 |
| 2. PMR and trunked radio systems | 46 |
| 3. Cordless communications | 82 |
| 4. Cellular radio systems | 107 |
| 5. Personal communication networks | 148 |
| 6. Communication satellite systems | 175 |
| 7. Acronyms | 226 |
| Index | 270 |

# Preface

Mobile communications is one of the fastest growing sectors of telecommunications, and the one in which there have been the greatest developments during the past few years. It spans a wide range of technologies from the simple radio pager to satellite based systems. This book reviews the many mobile communication technologies and applications in current use and those which are due to come into operation in the next century.

Chapter 1 describes radio paging, both on site and wide area. Chapter 2 introduces PMR and trunked systems, which have found a niche market in private communications, such as with the emergency services and fleet operators. This is followed in Chapter 3 by a description of cordless communication systems, from the long established domestic analogue receiver to the more advanced digital public CT2 and DECT systems.

Chapter 4 describes cellular radio, which represents perhaps the fastest growth area in mobile communications. This is followed in Chapter 5 by a description of personal communication networks (PCN) which may be considered to be a microcellular version of the main cellular system. The final chapter looks at the principles and components used in communication satellite systems, which currently promise to give global mobile communications into the next century.

Nine authors have contributed to this book, all specialists in their field, and the success of the book is largely due to their efforts. The book is also based on selected chapters which were first published in the much larger volume of the *Telecommunications Engineer's Reference Book*.

Fraidoon Mazda
Bishop's Stortford
April 1996

# List of contributors

**Malcolm Appleby**
MA
Cellnet
(Sections 3.1–3.4, 3.6)

**Fred Harrison**
BSc CEng MIEE
Cellnet
(Sections 3.3, 3.5. 3.7)

**C Lorek**
BSc (Hons) AMIEE
South Midlands Comms. Ltd
(Chapter 2)

**P J Marnick**
BSc (Hons) CEng MIEE
Hutchison Microtell Ltd
(Chapter 5)

**Fraidoon Mazda**
MPhil DFH CEng FIEE
Nortel Ltd
(Chapter 7)

**Richard J Munford**
CEng MIEE
Multitone Electronics plc
(Chapter 1)

**S C Pascall**
BSc PhD CEng MIEE
Commission of the European Communities
(Chapter 6)

**R G Russell**
BSc (Hons) CEng MIEE
Hutchison Microtel Ltd
(Chapter 5)

**R S Swain**
CEng MIEE
BT Laboratories
(Chapter 3)

# 1. Radio paging

## 1.1 Introduction

Radiopaging is a cost effective solution for staff location, alerting personnel and transmitting one way messages or data. Compared to land mobile radio, and cellular radio, radio pagers or 'bleepers' as they are known in hospitals, are much smaller and less expensive. They are also less expensive then CT2 or DECT handsets and are not limited in range as are these systems. With two notable system exceptions, however, the communication is only one way and without acknowledgement. Each pager has its own address and when the paging caller sends a coded transmission it alerts the bearer by means of an audible tone, a light and/or a vibration.

Paging systems fall into two classes called wide area and on site. The wide area user rents a paging receiver on a public service provider's network, while the on site user purchases his own private transmission system and paging receivers.

## 1.2 Markets

In Europe the industry has developed from single on site systems, the first of these being an inductive loop system installed in St. Thomas's hospital by Multitone Electronics in 1956. Transmissions were in the range 30kHz to 50kHz and each receiver was tuned to its own unique individual frequency. Loop current of 100mA around 200m to 500m length horizontal loops were driven by 5W to 10W drivers. Paging sensitivities of 100μA ft. sq. turn were typical. For a narrow tall building a vertical loop was used successfully. Many such systems are still in use, but the cost of installing the loop cable tends to be prohibitive.

2  Frequency allocations

National wide area systems in Europe, started by the Netherlands PTT in 1964, were followed by public member operator licences being granted to private consortia. In North America, the concept of private subscriber local wide area systems was developed early in the 1970s and by 1988 there were over 600 radio common units and 40 telephone companies providing services to 1.5 million subscribers.

## 1.3 Frequency allocations

Frequencies for on site and wide area paging vary by country. Reviews of the use of the spectrum have resulted in proposals for future change in some paging frequencies. The following is not exhaustive, but is indicative of the challenge facing the design engineers.

### 1.3.1 UK

1.3.1.1 *On site*

26.2375MHz to 26.8655MHz.
26.978MHZ to 27.262MHz;  31.725MHz; 31.75MHz; 31.775MHz (special allocation).
49.0MHz to 49.4875MHz.
49.425MHz; 49.4375MHz; 49.45MHZ; 49.4625MHz; 49.475MHz (speech permitted).
161.00MHz to 161.10MHz. Return speech (emergency/special licence only).
459.125MHz to 459.45 and 495.475MHz. (Shared with local communication systems.)
161.0MHz to 161.10 and 161.1125MHz. Return speech. (Shared with local communication systems.)

1.3.1.2 *Wide area*

**V.H.F.**

National and public systems exist on frequencies 138MHz to 141MHz and 153MHz to 153.5MHz. Private wide area schemes are assigned 'to manufacturers' preferred frequencies' in the 153MHz band. Additionally, the ERMES system occupies the band 169.425MHz to 169.80MHz. 'Overlay' and emergency systems are possible in conjunction with mobile radio systems, which cover the band 138MHz to 174MHz.

**U.H.F.**

Private wide area schemes are concentrated in the band 454.0125MHz to 454.825MHz. The 'Europage' (pan-European) scheme has been assigned to 466.075MHz. Again it is possible to have overlay systems in the mobile bands 450MHz to 470MHz.

### 1.3.2 Germany

#### 1.3.2.1 *On site*

27.51MHz; 40.76MHz and 40.68MHz. 20kHz channel spacing.
40.665MHz to 40.695MHz. 10kHz channel spacing.
468.35MHz; 468.375MHz; 468.4MHz to 469.150MHz.
Talkback 151.07MHz; 160.49MHz to 160.55MHz; 170.55MHZ to 170.79MHz.

#### 1.3.2.2 *Wide area*

Cityruf scheme 465.97MHz; 466.070MHz; 466.23MHz.
Private wide area, proposed frequency band 440MHz to 450MHz; 12.5kHz.

### 1.3.3 France

#### 1.3.3.1 *On site*

26.635MHz; 26.695MHz; 26.745MHz.
446.475MHz to 446.525MHz.
Talkback 152.0125MHz and 445.50MHz.
All frequencies are narrow channel (10/12.5kHz).

### 1.3.3.2 *Wide area*

31.30 MHz.
Pan-European 466.075MHz. Other overlay/emergency services schemes are operating across all the mobile bands.

## 1.3.4 Holland

### 1.3.4.1 *On site*

26.5, 26.6, 26.7, 26.8, 26.9MHz. (Low power systems.)
26.15MHz to 27.85MHz. (Excluding the above frequencies.)
39MHz to 40MHz.
Talkback 156MHz to 174MHz.
450MHz to 470MHz.

### 1.3.4.2 *Wide area*

PTT Semafoon scheme 154.9875MHz and 164.35MHz (Benelux).
Emergency services etc. across the mobile bands.

## 1.3.5 Italy

### 1.3.5.1 *On site*

26.20MHz; 26.35MHz and 26.50MHz.
459.65MHz and 469.65MHz.
Talkback 161MHz to 161.10MHz.

### 1.3.5.2 *Wide area*

A proposal was put forward for a PTT scheme on 161.175MHz, but this has been overtaken by the pan-European scheme on u.h.f.

## 1.3.6 Belgium

1.3.6.1 *On site*

Frequencies centred on 26MHz and 40MHz, and on 440MHz to 470MHz.

1.3.6.2 *Wide area*

PTT Semaphoon scheme 147.25MHz (national) and 164.35MHz (Benelux).
Overlay and emergency schemes on mobile frequencies.

## 1.3.7 Denmark

1.3.7.1 *On site*

29MHz to 31.42MHz h.f.
445.9MHz to 445.975MHz u.h.f.
Talkback frequencies at 146MHz, 161MHz and 422MHz.

1.3.7.2 *Wide area*

Public system (OPS) 469.5MHz; 469.65MHz and 469.95MHz.

## 1.3.8 Sweden

1.3.8.1 *On site*

26.1MHz to 26.958MHz.
Talkback 160MHz.

1.3.8.2 *Wide area*

169.425MHz to 170.0375MHz Public Regional Paging Service (MiniCall).

Overlay and emergency services etc. in the PMR bands at v.h.f. and u.h.f. 138MHz to 174MHz and 410MHz to 470MHz.

### 1.3.9 USA and Canada

The manner of frequency allocations in these countries is according to end user category, rather than specific frequency bands. This means that the whole mobile spectrum is encompassed. To cover both countries, pager manufacturers have to provide for 27MHz to 50 MHz h.f.; 138MHz to 174MHz v.h.f.; 445MHz to 470MHz u.h.f.. 900 MHz is also popular. None of the bands is divided into on site or wide area categories. 931MHz is used for satellite paging.

## 1.4 Paging receiver types

There are three types of paging receivers. The first and simplest is a tone receiver, which can have a number of different audible tone patterns (usually four or eight) to alert the bearer, but it does not display numbers or messages. The bearer takes a predetermined action e.g. telephone a given number, or proceed to a location. Since there can be up to eight tones others can indicate alternative actions or degree of urgency.

The numeric pager has a display of typically 12 or 24 digits thereby enabling the caller to send a telephone number that he wishes the bearer to contact. By prior arrangement the numbers sent can represent a message. The pager is usually able to store these numeric messages in a memory.

The third type of pager is the alphanumeric, which has a display of 16, 32, or 96 characters. Between 500 and 5000 characters of message can be stored in the memory and can be read by scrolling, although typically messages up to 30 characters are sent with information services requiring a 7K memory block. Unlike the tone and numeric pagers which can be called directly by telephone, the sender must use an alpha keyboard to send alpha messages. Generally wide area paging users send alpha messages via an operator at a bureau service while on site users receive alpha messages from the local paging system operator.

## 1.5 On site paging

### 1.5.1 Applications

On site paging systems are intended for the coverage of a premises or a local site completely within the user's control. These systems are usually owned and licensed by the user and vary from a single low power transmitter/control unit operating with five pagers to large high power, multi-site, multi-access systems with 1000 or more pagers.

All three types of pager can be used in on site applications but, because speech messages are also permitted, tone alert and numeric systems have generally satisfied the needs of staff location and one way message transmissions.

By preprogramming a series of fixed alpha messages into the receiver's memory, which are displayed on receipt of predetermined number sequences dialled by the calling party, a numeric pager provides many on site alpha numeric functions at low cost. With the growing demand for the transmission of data in the health care market and process industries, the on site alpha pager is becoming more in demand.

Sites vary from large warehouse areas with few personnel, where travelling time is saved by updating requirements from stores office to fork lift truck driver, to densely populated sites such as hospitals where emergency calls for groups of staff (cardiac arrest teams) can be summoned as a unit by the group call facility. A waitress, with a tone pager, can be notified by the chef when her next order is ready, and a broker can be continuously updated of selected stock prices when a few miles from his office, by using an alphanumeric pager on a private wide area system.

### 1.5.2 Code formats

There have been many different on site paging system manufacturers since 1956, and each has been free to devise its own code format. Early proprietary formats used sequential audio tone modulation using two inductor coils 2cm × 1.3cm × 1.3cm in the range of 500Hz to 3.4kHz. For example a tone of 640Hz had a coil of inductance 11.7H and a Q of 10. Vibrating mechanical reeds were also used to achieve higher Qs and more frequencies, thereby increasing system capacity, but at the expense of mechanical robustness. These were ultimately replaced by active filters which could then be made smaller with high Q and high stability by using hybrid thick film circuits.

By this time the EIA (Electronic Industries Association) had standardised on two tone and five tone sequential tone formats in the 67.0Hz to 1687.2Hz frequency range. Since such a two tone code can have a user capacity of over 3000 this format could be adapted for speech and was suitable for even the largest on site systems.

As the market demanded more features and facilities and even smaller receiver size, digital code formats were devised. Unfortunately the benefits, such as the narrow noise bandwidths of the two tone systems, meant a sensitivity loss, when using digital formats of 4dB to 6dB. A modern on site digital proprietary code format specification is shown in Table 1.1.

### 1.5.3 Small systems

A small modern paging system will probably consist of a transcoder and five tone or numeric pagers. The transcoder is a combined digital paging encoder consisting of a keyboard and a 10 digit display, and a small 1W transmitter incorporating an integral microphone. In the 25MHz to 54Mhz h.f. band a standalone base loaded whip aerial would be provided, but at u.h.f. a flexible stub antenna is connected directly into the encoder.

1W is sufficient r.f. power to provide good coverage in the office building environment, or 200 metres to 300 metres in a hotel block, and up to 8km on an open site. Range varies due to attenuation of buildings especially if steel reinforced structures or sun reflective windows are used. Attenuations at different frequencies can be encountered as shown in Table 1.2, which give average relative levels in dB. It has been found that, when walking with a pager, peak to trough variations of 30dB can be encountered due to Rayleigh fading effects, so this table is only a guide.

**Table 1.1** A typical digital proprietary format

| Code capacity addresses | 10 000 × 4 systems |
|---|---|
| Paging rate (call/sec) | Max 13 c/s tone alerts preamble. 40ch. 0.38 c/s. 80ch. 0.25 c/s |
| Data rate | 512bit/s |
| Message rate | 512bit/s |
| Radio channel | 25kHz |
| Modulation | ±4.5kHz FM FSK |
| Transmission mode | Manchester or NRZ |
| Multiple transmitter conditions | Quasi-Sync or sequential |
| Line transmission | Data modems 1200bit/s |
| Absolute line delay | 0.25ms + 0.195ms from transmitter. Telegraph distortion ≤10% Isochronous |
| Error correcting potential | 2 errors in 12 bits preamble. 2 bits in 32 non-message codewords. 1 bit in 32 message code words |
| Battery economy | Yes. Variable unframed |
| Message capability | No limit |
| Effect on receiver sensitivity | NRZ similar to CCIR code No.1. Manchester approx. 2dB worse |

At h.f., r.f. noise from electrical machinery, lighting, power lines and motor vehicles is prevalent. U.h.f. has marked benefits, in penetration of modern buildings and cellars. Some particularly r.f. noisy equipment used in hospitals e.g. diathermy units, are avoided for safety reasons by special frequency allocations for paging.

**Table 1.2** Attenuation at different frequencies

|  | 30MHz dB | 150MHz dB | 450MHz dB | 850MHz dB |
|---|---|---|---|---|
| Urban | +35 | +20 | +10 | +12 |
| Suburban | +20 | +10 | 0 | +2 |
| Rural | +6 | 0 | −10 | −8 |

## 1.5.4 Telephone coupling

For many users, with say 20 to 100 pager wearers, passing messages through a paging controller, who may well also be the switchboard operator, is inconvenient and direct paging access by the caller's own telephone extension is desirable. A telephone coupled encoder may be installed via one of the PABX extensions to satisfy this requirement. The units can have call transfer or absence registration programmed via the telephone for a limited number of pagers (e.g. 10). The system will be capable of sending tone, numeric messages, followed by up to 120 seconds of speech, but this is limited to 30 seconds speech transmission by regulations in the UK.

A 'meet me' facility is available via the PABX whereby the caller pages the person he wishes to contact, holds, and is connected directly through the PABX when the person dials the 'meet me' number.

Figure 1.1 shows the block diagram of a telephone coupled encoder which, via a telephone extension, decodes over dialled DTMF tones to generate paging calls. Programmed software is stored in a 8K EPROM. Two isolation barriers are provided, one to connect to a PABX and the other for a dedicated line. All 16 DTMF tones are decoded and encoded. Provision for different line levels has also to be made, e.g. −10dBm to −20dBm (600Ω) on the incoming telephone line, and attenuated to −13dBm (600Ω) out to the private wire.

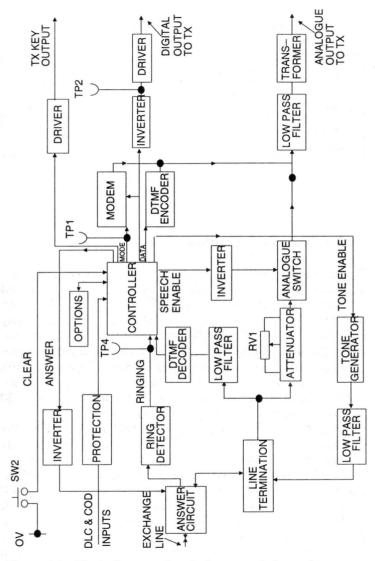

**Figure 1.1** Block diagram of a telephone coupled encoder

### 1.5.5 Regulating authority requirements

Duration of the speech part of paging messages is limited by many European countries to 30 seconds only.

In France this is limited to 10 seconds. While the usual compliance with and approval to radio regulations are required, telephone coupling equipment will also have to be approved by telephone line and safety authorities. In the UK BABT (British Approvals Board for Telecommunications) is responsible for type approval of telephone line connection equipment and for safety compliance to BS6301. In Germany the line approval is undertaken by ZZF. In the USA FCC has responsibility as does the Department of Communications in Canada.

### 1.5.6 Talkback paging

Portable units comprising a paging receiver and a transmitter capable of speech acknowledgement are used on about 10% of all on site systems. Such systems allow two-way speech and two-way signalling in any combination provided that CTCSS (Continuous Tone Controlled Squelch System) or selective calling are used in both directions. A base receiver is required fitted with a matching CTCSS or selective calling squelch.

In the UK this system is licensed as local communications and pairs of frequencies are allocated with the receivers in the 459MHz band, and transmitters in the 161MHz band. Typical specifications of a range of talkback pagers which are required for different European markets are shown in Table 1.3.

In France talkback paging is at 445.5MHz with a channel spacing of 12.5kHz and maximum ERP of 50mW.

An additional feature is that the control station can be switched to talkthrough, so that one receiver can talk direct to another. This is a useful facility for first aid teams or security guards.

The European Selective Paging Association (ESPA) are currently considering putting forward proposals to ETSI to replace the cross band talkback frequencies by 10 additional frequencies at u.h.f.

**Table 1.3** Typical talkback pager specification

| Receiver frequency range | 25MHz to 54MHz | 406MHz to 490MHz |
|---|---|---|
| Channel spacing | 10, 12.5, 20, 25kHz | 20kHz to 25kHz |
| Sensitivity (typical call) | 20µV/m | 15µV/m |
| Speech (12dB Sinad) | 40µV/m | 30µV/m |
| Adjacent channel select (typical) | 65dB | 70dB |
| Spurious response (typical) | 50dB | 60dB |
| Spurious emissions (max.) | 2nW | 2nW |
| Transmitter frequency range | 146MHz to 174MHz | |
| Channel spacing | 10, 12.5, 20, or 25kHz | |
| Output power (typical) | 20mW ERP | |
| Talkback time limits | Unrestricted/restricted according to regulations | |
| Audio distortion | ≤ 5% | |
| Spurious emissions (max.) | 0.25µW | |
| CTCSS (optional) | To EIA standards | |
| RS220 frequencies | 103.5, 114.8, 151.4, 167.9, 225.7, 250.3Hz etc. | |

## 1.6 Large on site systems

Figure 1.2 is an example of a large on site paging scheme and is fully featured. It has a remotely linked section of the system as would be found on a two sited hospital complex or a large brewery. Modular systems such as these may have as few as 15 or as many as 1500 pagers.

The common bus connecting these units forms a local area network with a bus cable length of up to 1km. The bus cable is 8 way, with one pair of data lines (RS485 standard), one pair outgoing speech, one pair return speech and the remainder used for earth connection. The data rate is 19.2kbit/s and digital data voltages are 0 and +5V. For the speech pairs frequency band is 300Hz to 3kHz at a line level of $-13$dBm.

The master/slave architecture, where one module is used as the master and the rest of the modules are treated as slave devices, operates a communications protocol using a token passing scheme. This scheme gives permission for slave devices to use the system data bus on command of the master. The master also holds a device parameter table, which stores the initialisation parameters for every module on its bus.

Operators can send paging calls and enter and update information by means of small local control units, or larger alpha numeric control units and, if on a large site connected by private wire, operation can also be made by remote control unit via a gateway module.

The telephone interface units shown are via interface cards, which provide a direct hard wired link between the paging system and PTT or other approved private lines, selector level circuits, extension circuits or tie line type circuits. The cards provide the correct electrical signal conditioning and safety requirements of such connections.

For direct machine driven paging calls, or push button driven paging calls, a remote contact interface is used. Typical uses are for callers who require entry to a locked building patrolled by a security guard carrying a pager, or by process engineers who are supervising many pieces of continuous running equipment, and need to be notified of faults, or important machine parameter changes.

Radio paging 15

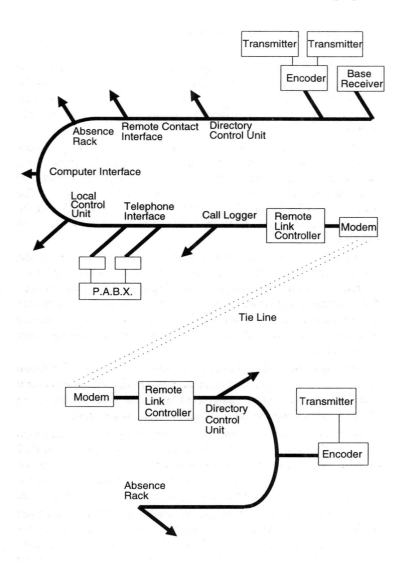

**Figure 1.2** A fully featured on site paging system

Connection to a standard computer by RS232c line is shown. The computer operator could also page by directory viewing if this software is entered into computer memory. The call logging facility provides a record of times and user numbers actually called on the system, which is useful when reviewing system usage.

Absence rack units automatically register the user as absent, when he places his pager in one of the slots. Subsequent callers will be informed that he is unavailable or the call may be transferred to an alternative pager. These are usually sited at staff exits and usually contain a charging facility to recharge pagers overnight.

The remote link controller allows the system data bus to be extended via any RS232c compatible link. For examples modems and telephone lines, KiloStream, MegaStream, or data networks can all be used to carry the link.

The system has a large data base containing user, remote contact interface and team records. This provides comprehensive data processing and organising functions for these records and the memory is protected against power supply failure by battery backup. The memory allows all records to be sent to a dedicated printer for a hard copy.

Through this unit, a pager's identity can be transferred to another pager, if the first pager is withdrawn from service for repair. If a user is going off site, for example, or if the user is on holiday, he can have his calls transferred to another user's pager, by entering the relevant data and identity into the data base. New teams of users can be constructed readily by the operator, for group calling purposes.

Figure 1.2 shows paging transmitters from each of the two sites and also a base receiver, to enable talkback paging to be used.

## 1.6.1 On site transmitters

On small systems 0.5W at u.h.f. and 1.0W at h.f., are sufficient transmitter r.f. powers to satisfy the small site application, and these can be integral with the encoder. For larger sites, higher powers are required with appropriately sited antennas. A limit of 5W ERP is generally imposed by regulation in Europe, (limit is 2W in Belgium) so it often becomes necessary to install more than one transmitter on the large area sites and campus (100 sq. km). Table 1.4 shows typical

**Table 1.4** Typical on site transmitter specifications

| Frequency range | 25 to 54MHz | | 390 - 490MHz | |
|---|---|---|---|---|
| Frequency stability | ±10ppm from −10 to +55°C | | 5ppm from −10 to +55°C | |
| Output power | 5 watts nominal | | 5 watts nominal | |
| RF load | 50Ω nominal impedance, VSWR better than 2.1 | | 50Ω nominal impedance, VSWR better than 2.1 | |
| Channel spacing | 10/12.5/20/25kHz | | 20/25kHz, 12.5kHz | |
| Modulation type | FM | FM with FSK | FM | FM with FSK |
| Modulation system | Data F2D; Speech F3E | Data F2D, F1D; Speech F3E | Data F2D; Speech F3E | Data F2D, F1D; Speech F3E |
| Spurious emissions | Better than 73dB (250nW) below carrier. 100kHz to 2GHz (excluding wanted and two adjacent channels) | | | |
| Line input | −33dBm to −13dBm 600Ω balanced line | | | |
| Mains input | 100V to 240V a.c. nominal 50/60Hz | | | |
| External transmitter control | −13dBm 600Ω speech and data outputs for modulation/control of a second transmitter | | | |
| Back channel input | TTL compatible for line transmission of data | | | |
| Line synchronisation (P211 only) | Input/output for future installation of optional modules | | | |
| Master-slave | Connectors provided for optional | | | |
| D.C. output | 13.5V d.c. 50mA maximum for ancillary equipment | | | |
| Panel display (7 segments) | Standby transmitter keyed and transmitter failed | | | |
| Internal LED indictors | Phase-lock fail high reverse power and line sync fail | | | |
| Alarm outputs | Open-collector logic outputs for transmitter keyed (on/off) and r.f. output (absent/present) | | | |
| Operating temperature | −10 to +55°C | | | |
| Humidity | Max 90% R.H. non-condensing at 40°C | | | |

on site transmitter specifications. Greater r.f. powers, up to 25W or 100W depending on the system, are allowed in North America.

Since a variety of proprietary analogue and digital code formats, as well as speech, have to be transmitted, modulation techniques differ. FM is suitable for many applications but for non return to zero codes such as ITU-R RPC1 (POCSAG), FSK modulation is also required.

Although POCSAG is a large capacity code not supporting speech, it can be used on site. Mixed paging code formats, both analogue and digital, can be found on large on site systems.

## 1.6.2 Transmitter synchronisation

On large sites, or in areas of difficult propagation, it is often necessary to use multiple transmitters to give adequate radio coverage. Sequential polling is the obvious solution for those on site systems where speech is not used.

When both paging calling and speech is used either quasi synchronisation or full synchronisation may be used. The former requires transmitters of high inherent stability, $\pm 2 \times 10^8$ over $-10^{\circ}$C to $+50^{\circ}$C, and hence is expensive for on site use. Full synchronisation can give poorer radio performance in that a fixed standing wave pattern is set up, leading to fixed null areas. The cost is however less for small systems and thus line synchronisation units are used as one method of fully synchronising transmitters.

Transmitters in both the h.f. and u.h.f. bands will probably use phase locked techniques with a crystal reference oscillator as the frequency determining element. An audio signal in the 2.8kHz to 3.0kHz range, derived by dividing this oscillator, may be sent down a telephone line. At the slave transmitter, the crystal oscillator, divided to the same frequency, can be fed to a phase detector with the filtered signal from the line. The resultant error signal then controls the slave oscillator frequency using varactor diodes.

The telephone line interfaces from the master and to the slave will be required to meet the national standard of the country in which the equipment is used. Some degradation in the S/N ratio (modulation on/modulation off) will occur from noise on the synchronisation line.

Since the effective multiplication of the controlled crystal is much greater in the h.f. band than in the u.h.f. band, a specification of signal to noise ratio of 40dB at u.h.f. requires a much better signal to noise of the crystal oscillator itself, than is needed to meet 40dB in the lower frequency h.f. case. It should be such that noise on the line will not interfere with normal working with a 20dB signal to noise ratio at the slave end of the line. Noise bursts however will temporarily unlock the synchronisation.

A guide to a specification is that the slave must synchronise correctly with noise bursts of maximum duration of 1 second, at an amplitude equal to the synchronising signal. Frequency of the master oscillator, may drift by up 40ppm at h.f. and ±10ppm at u.h.f. over worst conditions of temperature and/or time. Acquisition of lock and synchronisation of performance should be met over the whole of these ranges. Speech signals may be passed down the synchronisation line, in which case a low pass filter will be used to attenuate speech frequencies above 2.5kHz. This leaves the range 2.8kHz to 3.0kHz to be available for the control signal.

Another form of synchronisation uses a coaxial cable to pass the master oscillator frequency signal to the slave transmitter multiplier chain. For distances over 100 metres an additional amplifier will probably be necessary.

# 1.7 Wide area paging

## 1.7.1 Tone code formats

City wide paging was developed by US BELL in the early 1960s, followed by the Dutch PTT who opened their national Semafoon network at v.h.f. using Motorola EIA 5 Tone signalling. This had an address capacity of 100,000 with a calling rate of 5 calls per second. Tone duration was 33mS and tone frequency relationship was an arithmetic progression.

The first pan-European system, Eurosignal, commenced in 1967, uses 6 tone sequential system and has a capacity of over one million, with a calling rate of 1.25 calls per second. Tone duration is 100mS and tone frequency relationship is a geometric progression.

## 1.7.2 Digital code formats

Bell Canada introduced its wide area digital paging system in 1970. Compared to sequential tone formats, digital formats can provide greater system address capacities, much greater battery economy and thereby use smaller power cells, faster call rates, pager size reduction by integration, additional functions such as storage of calls and messages in memory, multiple addresses and many other features. Sweden inaugurated its national digital paging service using a subcarrier on the national broadcast system in 1970 and Japan, already up to 600,000 on a tone system opened the NTT digital system in 1978. Meanwhile in North America the majority of systems supplied since 1973 were based on the Golay code format.

In 1978 in the UK the British Post Office announced and agreed a common digital paging system, POCSAG (Post Office Code Standardisation Advisory Group). This was the result of twp years of work by the British Post Office and representatives from 16 major pager manufacturers under the chairmanship of Mr R Tridgell. This standard was adopted by the ITU-R as Radio Paging Code RPC No.1. In 1980 the advisory group met again with world wide representation, and minimum standards for numeric and alpha numeric messaging were agreed. Table 1.5 compares the basis parameters of three of the most significant digital paging code formats used to date.

A new generation of high speed digital code formats have been formulated. These are ERMES, which is descried later, FLEX, which has been developed by Motorola Inc., and APOC, developed by Phillips Telecom. Binary FSK POCSAG at 2400bits/s can provide 32k users per channel. ERMES using 4 level FSK, 6250bits/s can provide 145k users per channel. FLEX using 4 level FSK at 6400bits/s can provide 157k users per channel. APOC using 4-pam/fm modulation at 6400bits/s can provide 395k users per channel. All these capacities are for 40 character alphanumeric messages.

## 1.7.3 POCSAG

An outline of the POCSAG format is shown in Figure 1.3. A transmission sequence starts with a series of alternate digits, 101010..... as

**Table 1.5** Comparison of digital code formats

| Parameter | GSC | POCSAG | NTT |
|---|---|---|---|
| Code type | Golay 32.12 | BCH 32.21 | BCH 31.16 |
| Address capacity | 1 × 4 address | 2 × 4 address | 30k × 2 address |
| Tone call rate | 5 call/s | 15 call/s at 512 baud | 5.7 call/s |
| Message call rate | 2.5 call/s 12 CH; 0.45 call/s 80 CH | 5 call/s 10 CH; 0.52 call/s 80 CH | |
| Transmission rate | 300/600 | 512 or 1200 | 200 units |
| Decoder type | Asynchronous | Bit synchronisation and word framing | |
| Random errors: Detection | 6 | 5 | 6 |
| Random errors: Correction | 3 | 2 | 3 |
| Burst errors: Detection | 11 | 11 | 15 |
| Burst errors: Correction | 5 | 5 | 7 |

a preamble to awaken the paging receiver to obtain synchronisation. This lasts for at least 576 bits which is equal to the duration of one batch plus one codeword. A synchronisation code word is then followed by eight frames, each two code words long. Each pager is allocated an address in one of the eight frames and having gained synchronisation it need not turn fully on until that frame's time period, thereby saving battery power.

## 22 Wide area paging

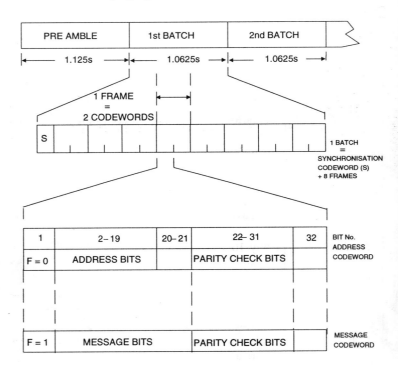

**Figure 1.3** RPC No. 1 POCSAG code format

If there is a message for a pager then it immediately follows the address code word. It can be any number of consecutive code words and may continue for more than one batch, but the synchronisation code word remains at the front of every batch. The message end is always followed by the next address code word or an idle word.

The receiver distinguishes address from message code words by reading the first bit (F) which is a flag bit, equal to zero for an address code word and equal to one for a message code word.

Address code words consist of 21 bits the 18 most significant of which determine the address. The three least significant are not transmitted, but are used to identify the frame in which the address code word is transmitted. Bits 20 and 21 are two function bits which serve to indicate which of the 4 addresses of the pager is being called. Both address and message code words have a final bit chosen to give even parity following the parity check bits.

For numeric messages only a 4 bit per character format is used, restricting messages to decimal numerals, U (urgency indicator), space, hyphen, opening and closing brackets and a spare symbol. This format saves air time, but for full alphanumeric messages a full 150 7-bit character set is used. When using the numeric only format the pager address with the message will have its function bits set to 00, but when the seven bits per character format is used the pager address has its function bits set to 11.

### 1.7.4 Wide area transmission

Transmitters of typically 100W/250W are used to cover wide area paging zones, but there are many exceptions of much higher powers being used across the world depending on local terrain. The coverage area for the system is generally divided into zones. These zones vary considerably in size from 256 sq. km to 25,600 sq. km depending on service that is being provided. The principle however is the same, in that the paging subscriber pays for the use of his pager in chosen zones. All paging calls are usually sent to a control centre where they are dispatched to the relevant zones transmitter controller. The paging calls are stored, batched and then transmitted. Some systems transmit calls twice or even three times depending on service provider.

Transmission in these zones can be sequential or simultaneous. Often only one frequency is employed but BT, who claim to cover 97% of the UK population with an estimated half a million subscribers, have 2 frequency channels and 40 zones. Since zones are adjacent, with varying degrees of signal overlap and co-channel interference, calls are transmitted in time slots. Zones are allocated a time slot such that overlap is minimised. Time slots can be varied in length depending on the paging population in each zone.

For the RPC No.1. code format sequential transmission can also be used within a zone, but where traffic is heavy simultaneous transmission must be employed. Since the transmitters are not locked together they run in a quasi synchronous mode, and therefore require higher stability with temperature and ageing ($2 \times 10^8$ over $-10°C$ to $+50°C$ and 0.5 ppm /year) than a transmitter used for sequential transmission. They must have a maximum line delay differential no greater than 250μs and isochronous distortion no greater than 10%.

It is possible in zone overlap areas or with time slotting, to receive the same paging call twice, albeit one of the calls is at a much lower signal strength level than the other. Many pagers have memory configured such that they reject the second identical call.

#### 1.7.4.1 *Digital modulation*

To modulate the digital signal phase modulation is unsuitable, and therefore Frequency Shift Keying (FSK) is used. For the particular characteristics of POCSAG and similar formats, non return to zero FSK is used. This means that the modulation frequency is either +4.5kHz or −4.5kHz and the modulation does not diminish to zero during the change. Three common types of digital modulation for a binary number are shown in Figure 1.4. In NRZ level a one is represented by one level and a zero by the other level. In NRZ mark, a one is represented by a change in level and a zero by no change in level. In biphase level a one and a zero are represented by a positive/negative pulse sequence, each pulse being half a symbol wide.

### 1.7.5 Swedish Mobilsoekring system (MBS)

This nation wide system was launched by the Swedish PTT in late 1978 and uses the existing broadcast FM infrastructure. It is a cost effective system for thinly populated countries; 8 million people in 449,750 sq. km. Use is made of one of the stereo broadcast programmes in the 87MHz to 104MHz band, which covers most of Sweden and some of Norway and Denmark.

Transmission is by the differentially coded binary information modulating a 1.187 kHz (±0.1Hz) tone which itself modulates a 57kHz (±6Hz) sub-carrier with maximum deviation of ±3kHz. This 57kHz signal is transmitted as a sub-carrier with the FM programme signal. The sub-carrier is phase locked to the 19kHz stereo multiplexed pilot signal.

Radio paging 25

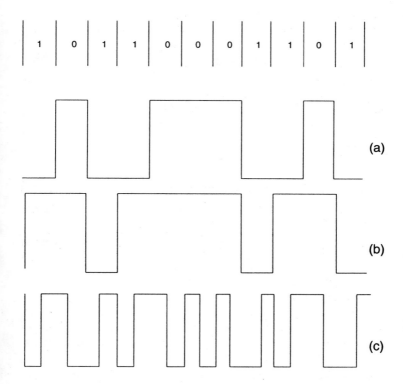

**Figure 1.4** Three common types of modulation used for digital paging codes: (a) NRZ level; (b) NRZ mark (differential encoding); (c) Bi-Phase level (Manchester code)

Because the programme operates in different areas on different frequencies the receiver scans the 87MHz to 104MHz band every 10 seconds looking for an MBS system identification code. The caller dials a four digit entry code, a subscriber 6 digit number and further digits for options, such as a privacy code which is used to prevent unauthorised callers paging the subscriber. A 52 bit paging code is sent which consists of two blocks of 16 information bits and 10 parity check bits.

## 1.8 Paging receiver design

The challenge to the paging receiver designer is always to package more facilities into a smaller unit without losing radio performance. In 1970 a typical size was 13cm × 6cm × 2cm but today wide area tone units can be 5cm × 4cm × 1.5cm, or 5cm × 1.5cm × 1.5cm depending on the type of battery used. A numeric pager the size of a fountain pen has been used for several years, and a pager in a wrist watch has now been launched.

Credit card style pagers 7.5cm × 0.5cm × 5cm are currently available on the Hong Kong national network at 280MHz and use a 1.4V zinc air cell of 700mAh capacity. Many of these more ambitious physical formats have compromised sensitivity when actually worn on the body.

Paging receiver designs must match world wide requirements in the range 25MHz to 900MHz.

### 1.8.1 Receiver specifications

Typical receiver specifications are shown in Table 1.6. Each country has its own regulations set by its type approval authorities (PTT's, FCC etc.) and they usually differ.

Under the European Community Directives, harmonisation of specifications will occur and if a paging equipment is awarded the CE mark then it will be approved for sale in any EEC market. This will save the costs of multiple approvals and reduce time delays from the manufacturers.

The regulations appertaining to paging receivers both in Europe and North America will be restricted to EMC interference from the receiver and, in Europe, immunity from external interference to the receiver. The European regulation also specifies a minimum immunity to electrostatic discharge.

Operators of wide area systems will however still determine their own more comprehensive minimum performance requirements specification, which the manufacturer's equipment will have to meet if he wishes to sell to them.

Radio paging 27

**Table 1.6** Typical receiver specifications. On body sensitivity is dependent on code format and pager usage whilst co-channel rejection depends on code, speech and measurement techniques

| Parameter | 25 to 49MHz | 138 to 174MHz | 406 to 470MHz |
|---|---|---|---|
| On body sensitivity: face-on | 8μV/m | 5μV/m | 8V/m |
| On body sensitivity: 8 pos.ave | 12μV/m | 8μV/m | 10μV/m |
| Channel spacing | 10/12.5kHz or 20/25kHz | 25kHz | 20/25kHz or 12.5kHz |
| Adjacent channel selectivity | >50dB or >60dB | >65dB | >65dB |
| Spurious emissions up to 1GHz | 2nW | 2nW | 2nW |
| Co-channel rejection | −2dB to −10dB | −2dB to −10dB | −2dB to −10dB |
| Inter modulation rejection | >45dB | >50dB | >50dB |

## 1.8.2 Receiver architecture

The receiver architectures that are usually employed are single superhet, double superhet and direct conversion. The latter lends itself to a high degree of integration to two or even a single multi chip circuit, plus supporting circuitry of voltage multipliers, integral antenna, alerter transducer and battery. This arrangement makes for a very small pager especially if a button cell is used as the battery. Initially it was found to be vulnerable to interference from nearby high power transmitters due to poor AM rejection, but now there are several techniques for minimising this effect.

## 28 Paging receiver design

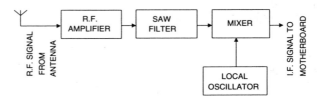

**Figure 1.5** Superhet r.f. block diagram

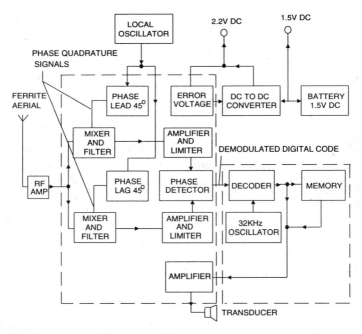

**Figure 1.6** Direct conversion receivers

Analogue processed speech is not possible using direct conversion thus it is only suitable for wide area applications. Typical single superhet and direct conversion block diagrams are shown in Figure 1.5 and Figure 1.6. Figure 1.7 shows the block diagram of the decoder and audio sections. A wide area decoder would omit the audio ampli-

Radio paging 29

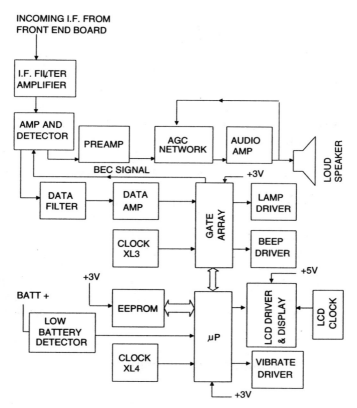

**Figure 1.7** Decoder and audio block diagram

fier section. Speech is shown outputting to a speaker but it may also be simply channelled through the beep transducer.

The latter allows a smaller on site pager but many customers insist on better quality and louder speech than is possible from the small beep transducer, whose audio characteristic will have a substantial resonant peak in the 2.5kHz to 3kHz band. This peak is necessary to gain maximum loudness from the transducer at the beep frequency. Transducer beep output is often a requirement in the operator's specification, usually in the range 75dB to 85dB SPL measured in an anechoic chamber facing the transducer output, and at a point 30cm away.

The front end r.f. noise factor using the very low voltage supply rail and minimum current, is a very important factor in the S/N ratio which finally enters the decoder, but the inefficient antennae that have to be employed largely determine the overall sensitivity.

### 1.8.3 Antennas

Ferrite antennas are usually used for 25MHz to 54MHz band receivers and sometimes in the 137MHz to 174MHz band, and are wound with flat copper strip. For the 430MHz to 490MHz band antennas can be a helical coil, discontinuous directional ring radiator, etc. Unfortunately, because of the limited space within the pager and close proximity of other components and earth tracks, u.h.f. antennas are not able to operate at their optimum. While antennas may result in good calling sensitivity (e.g. 5µV/m) measured in an open field attached to a pole or on a strip line in the lab, performance maybe considerably reduced when worn by the user due to the effect of the body.

### 1.8.4 Performance measurement methods

There are two particular difficulties in performing measurements on the paging receiver:

1. The calling signal must be the pager's digitally modulated address.
2. The integral antenna prevents direct connection from the signal generator.

Not withstanding the difficulties, the measurement method which has been the most commonly used is to set the signal generator below the level of calling and increase this level in 0.5dB stages, sending the pager's address 5 times at each stage. The generator level, at which 5 successful calls out of 5 calls are sent, is deemed to be the sensitivity.

This can be shown to be related to a theoretical 87% probability of calling.

Another method, from IEC No. 49, and mandatory for pagers, when determining the sensitivity of equipment used on the German Cityruf system, is as follows. The signal generator level is adjusted so that less than 10% of the tone only calls are successful. The signal is continuously repeated so as to ascertain each time whether a tone only call is successful or not. The level is increased by 1dB until three successive calls are successful. The level is then noted. The level is reduced by 1dB and the new level noted. The signal is then sent 20 times. The level of the signal:

1. Remains the same if a tone only call is successful.
2. Is increased by 1dB, and the new level noted if a tone only call is not successful.
3. Is reduced by 1dB and the new level noted if 3 successive tone only calls are successful.

The pager sensitivity is deemed to be the arithmetic mean of the noted levels.

As direct line connection of the modulated call is not possible, the pager must be placed in a uniform field strength capable of being varied. Three methods are commonly used:

1. The pager is strapped to a wooden pole facing the transmitting antenna 1.5m up from the ground in an open field, and is sent a signal from a calibrated antenna 30 metres away. It is important that the area is clear of trees and nearby buildings, to avoid reflections distorting the uniform electromagnetic field.
2. The pager may be placed in the centre of a strip line.
3. The pager is placed in a TEM (Transverse ElectroMagnetic) cell.

The first method provides an absolute result directly, but due to the nature of the calling process, several results should be averaged. Although the strip line and TEM cell fields may be calculated, it is usual to take the portable TEM cell out to the free field site, and correlate the face-on level determined on pole, directly to the required TEM cell level.

The second and third methods require a radio noise free environment which, along with portability, is the main advantage of using a TEM cell as in the third method.

### 1.8.5 Free field sensitivity

Because, in practice, transmissions may arrive at a receiver from any angle, the sensitivity specification is very often expressed in terms of the free field 8 position log average. For this purpose the pager on pole is rotated and measured every $45°$ starting with E1. The sensitivity is said to be $E_m$, given by Equation 1.1.

$$E_m = \left( \frac{8}{\frac{1}{E_1^2} + \frac{1}{E_2^2} + \ldots + \frac{1}{E_8^2}} \right)^{1/2} \tag{1.1}$$

The relationship between face-on free field sensitivity and the 8 position average is fixed for a given receiver design, but needs to be determined by field measurement (a loop aerial will have significant nulls at $90°$ and $270°$ with respect to face-on, whereas a helical mono-pole will have a much more circular polar diagram). Thus a correlation factor between the 8 position free field sensitivity and the face-on figure in the TEM cell can also be established.

For some public wide area operator's specifications, pager on pole measurements are called up. Others specify pager worn on body measurements in free field. While different antennas are affected in different ways by the body, those favoured are those whose performance is enhanced by the body. Since different bodies affect the same antenna to different degrees, in order to establish a claim for on body sensitivity, measurements of sensitivity of the same pager on 30–50 different people are required. However, there appears to be no obvious correlation between height or weight of the wearer and the sensitivity measured. It has been found that sensitivity measurements made on various pager designs, each worn by 40 people, can have standard deviations of approximately 4.0μV/m.

### 1.8.6 Power sources

To meet world wide markets the pager must be designed with easily available primary cells because, with the exception of on site systems with more than 20 to 40 pagers, the user generally does not wish to purchase a charger. This considerably restricts the choice of cell since suitable button cells are not freely available outside N. America and major European countries.

As a battery life of 800 to 900 hours or about 2 to 3 months, depending on paging receiver and system usage, is required then the AA cell is widely used. With improvements in circuit techniques and world wide availability, the AAA cell with renewal periods of 1 to 2 months is now often used. Rechargeable AA cells of 500 mAHrs capacity may be employed and single or multiway charging racks are available. Non rechargeable cells, such as zinc, alkaline, manganese dioxide, start life with 1.5V and slowly decrease to around 0.9V. Rechargeable cells only start at approximately 1.25V but maintain this level, then rapidly collapse when discharged. The paging circuitry has to take account of these different characteristics especially at the lowest working temperature of $-10^{o}C$.

### 1.8.7 Battery economy

To provide acceptable primary cell life times suitable paging code formats and receiver designs are required. Referring to Figure 1.8 and assuming the code is POCSAG, then the decoder circuit always has power on, but only switches on the radio circuit every 0.4 to 1 second (depending on data rate) to search for preamble. If preamble is detected a sync word search commences while power is still maintained to the receiver.

When a sync word is found the pager synchronises with the incoming data stream and immediately turns off the radio circuits until the pager's own time frame in the batch of 8 is approaching. The radio is then turned on for say 10ms settling period followed by a 64 bit period to read the address code words. If its address is found then a tone pager will beep and flash, and a numeric or alpha pager will decode and display and/or store the message, before the radio section is once again switched off.

## 1.8.8 Receiver programming

Each receiver needs to be programmed with its individual address or addresses and with its feature options. This is carried out by a programming unit which can be used to programme the receiver by the manufacturer, and to reprogramme it by the customer service engineer if desired.

Each pager has a type of plug in programmable memory. In the early 1980s this may well have been a bipolar array such as an $8 \times 4$ diode matrix with fusible links. Programming would take place by passing currents of 1A through these links for a period of 10ms to form an open circuit. The disadvantage with this type of memory was that to change the programme a new memory had to be used to replace the old one.

The design of the programmer unit had to ensure that the current and time for breaking the links was carefully controlled otherwise it was possible for the fused links to reform. The operator programmed the pager by means of a keypad on the programmer unit.

As personal computers became less expensive and more readily available, field programmers for service work stations were designed using a simple driver interface between the PC and the pager memory. As EEPROMs became available these were used as the programmable pager memory.

Early versions of 128 bits had sufficient memory for pagers with 2 addresses 31 bits long, and a series of programmable options. These were programmed by a PC controlled device but needed 15V to satisfactorily programme the EEPROM. Later versions of the EEPROM with 4K of memory and 3V read/write capability superseded the earlier types and made possible the storage of precoded alpha text, different language texts and customised characters. These options are incorporated in the pager memory and are called up by transmitting a numeric code. For alphanumeric pagers which require greater memory storage for received messages, static RAMs working from 5V and capable of storing 6000 characters are used, the disadvantage with these devices is that they need separate battery backup.

For numeric pagers which are only required to store short numeric messages the RAM of the microprocessor can be used without battery backup. The advantage of 3V read/write EEPROM is that off air programming is possible where the pager receives reprogramming information transmitted to it from the base station. The low voltage means it can rewrite its own memory.

## 1.9 Private wide area paging

### 1.9.1 Overlay paging

Overlay paging is a system where paging addresses are sent over a PMR two way vehicular radio system to directly alert the driver, who has temporarily left his vehicle, to the fact that a message awaits his attention. Such additions to PMR schemes are permitted, though in the UK they are subject to the following conditions:

1. The number of mobile paging receivers must not exceed the number of mobiles in a system and each paging receiver must be used only in conjunction with a particular mobile station.
2. The normal base station transmitter power may not be increased.
3. The use of additional stations cannot be licensed solely on the grounds that they are necessary for overlay paging.
4. The paging system must use the same type of modulation as that of the mobile system in which it is employed.

### 1.9.2 Revertive paging or secondary calling

Revertive paging, or secondary calling, is an addition to PMR systems where paging addresses originating at the base station are received by the vehicle radio and retransmitted to a paging receiver carried by the driver.

### 1.9.3 Private off site paging

These are non public subscriber large area (up to 10 sq. miles) systems. To avoid interference between neighbouring small on site systems, alternative frequencies are chosen where available. In the UK, in order to minimise interference, manufacturers who are members of the Radio Paging Association, determine the appropriate channel by entering the ordnance survey coordinates of the transmitter location into a data base of existing systems.

It is possible though, for a group of adjacent sites to timeshare the same channel frequency. In the UK specific time slots of 15 seconds in a 60 second cycle are allocated to a site to transmit its paging calls. Code formats are proprietary and speech is not allowed. In the UK 5 frequencies of 25kHz channels in the band 153.375MHz to 153.475MHz are allocated and maximum transmitter ERP of 25W is allowed.

### 1.9.4 German POSP system

A new u.h.f. private off site paging system (POSP) has been proposed in Germany and the specification has been published. This is not an ETSI initiative, and currently only applies to Germany. Each site's transmitter is allocated a specific permanent 6 second slot in a 60 second cycle.

The time slot sequence is standardised (Numbered 1–10) and begins with time slot No. 1. at the start of each new minute. The allocation of the time slot is carried out by the Deutches Bundesposte, when the licence is issued. The use of more than one 15W transmitter synchronised to the same time slot is allowed, in order to improve transmission distance within a system. A sequence of time slots can be allocated to one user where the system has a high call rate. The ZZF issue a range of address codes to the manufacturer upon equipment approval.

**Table 1.7** ITU-T radiopaging code No. 1

| Parameter | Value |
|---|---|
| Frequency | 466.075MHz ±50Hz (frequency offsets max. ±1800Hz allowable) |
| Deviation | ±4.0 or ±4.5kHz depending on country |
| Channel spacing | 20 or 25kHz depending on country |
| Transmission rate | 1200 ±10 Baud |
| Signal rise time | ≤275µs |
| Character set | Special characters e.g. umlaut ($\ddot{a}$, $\ddot{o}$) are both possible and permitted and will be specified by each country involved |

## 1.10 Pan-European systems

### 1.10.1 Euromessage

This system (Table 1.7) opened in late 1989 and is now operational in major towns and industrial regions in France and West Germany, in the UK in London and Home Counties, and some parts of Italy. Full messaging is possible in each individual country and the common specification is shown below. Additional features are added according to the National Operator's requirements.

The disadvantage of this system is that since it utilises existing paging networks in each country, it is only possible to call a pager somewhere in Europe through its own national base.

1.10.1.1 *Transmission*

Time slotting of transmitters is used to ensure decoupling of adjacent simulcast areas. One, two or a maximum of three simulcast areas form a paging zone depending on traffic and country. The duration of the time slot cycle can be adjusted to conform to traffic requirements

**Table 1.8** Methods used for Euromessage

| Country | Germany | U.K. | France | Italy |
|---|---|---|---|---|
| Preamble length in bits | 576 | 608 | 576 | 576 |
| Synchronisation codeword | RPC1 | RPC1 | RPC1 | RPC1 |
| Idle codeword | RPC1 | RPC1 | RPC1 | RPC1 |
| Max batches between preambles | 24 (or 42-96) | 61 | 25 | 25 |
| Transmission cycle time in seconds | 84 (or 66-141) | 30 | 11.8 | 12 |
| Number of timeslots | 3 or 1 | 1 | 1 | 1 |
| Longest timeslot in seconds | 28 (or 22-47) | 30 | 11.8 | 12 or 24 |

and the differences are shown below. In the case of Germany the fixed figure refers to the standard Cityruf system and the range allows the possibility of expansion for Euromessage as traffic increases.

The methods of filling a time slot with idle words under low traffic level varies between the 4 operators (Table 1.8) as does the total length of time slots.

### 1.10.1.2 *Euromessage receiver*

Numeric receivers must be capable of receiving at least 15 characters per call and alphanumeric radio paging receivers at least 80 characters per call. The remaining basic specification for the receiver is as in Table 1.9.

## 1.10.2 ERMES

The ERMES standards are the result of an ETSI Working Group. All operators use the channels within the 16 allocated, and the same

**Table 1.9** Euromessage receiver specification

| Parameter | Value |
|---|---|
| Minimum sensitivity: on body tone only (salty man) | ≤ 29µV/m |
| Minimum sensitivity: tone only free field | ≤ 32µV/m |
| Adjacent channel selectivity | ≥ 60dB |
| Co-channel rejection | ≥ –8dB |
| Intermodulation response | ≥ 50dB |
| Spurious response rejection: image, above or below 47–60MHz 87–108MHz 174–230MHz or 470–862Mhz | ≥ 50*dB* |
| All other cases | ≥ 60dB |

protocol, so that one operator's pager will work on another operator's system.

Thirty three operators in countries including the UK, France, Germany and Italy, signed a Memorandum of Understanding, to provide a system that allows international roaming, and would serve each of their capital cities and 25% of total service population (90 million people) by December 1993. The pager will scan 16 v.h.f. paging channels, many of which are already used in these countries for other services.

1.10.2.1 *Type of service and specification*

Services to be provided are: tone, numeric, alpha numeric and transparent data. Additional optional features are related to: paging acknowledgement, call destination, three levels of priority, charging services, protection against inadvertent message loss, call privacy and bureau services.

The specification for ERMES is given in Table 1.10.

## 40 Pan-European systems

**Table 1.10** Specification for ERMES

| Parameter | Value |
|---|---|
| Frequency range | 169.4125MHz to 169.8125MHz |
| Channels | 16 channels 25kHz spacing |
| Modulation | 4 level (4-PAM/FM) ±4687.5 and ±1562.5Hz |
| Data rate | 6.25Kbit/s |
| Symbol rate | 3.125kbaud |
| Error correction | 2 bits (30, 18) shortened cyclic code |
| Interleaving to provide burst error correction | Message only to a depth of 9 code words |

#### 1.10.2.2 *Code format*

The construction of the transmitted code is shown in Figure 1.8. A sixty minute sequence is shown, of which a 1 minute cycle and then a 12 second subsequence is detailed. Batch I is expanded into its synchronisation, systems information, pager addresses block, and pager messages block of code words.

Battery economy is built into the format at various levels, and to assist current saving, addresses and system information is not interleaved. The pager will be addressed in only one of the sixteen batches (A-P) in a subsequence, enabling it to ignore each of the other fifteen. Once a pager has been addressed within its batch, the corresponding message may be in any of the following batches of that subsequence, or in the following subsequence, up to a point 12 seconds from the start of the batch containing the initial address. The addressed pager must therefore monitor all the subsequent batches until it finds one with its own address appended, or it times out after 12 seconds. Battery economy is also possible in other ways, but the process becomes complicated.

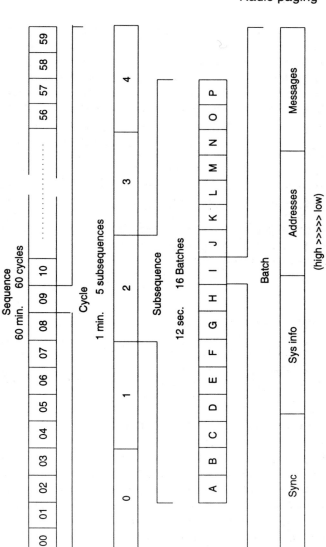

**Figure 1.8** ERMES protocol timing

## 42 Indirect satellite paging

In multi-frequency networks the paging receiver will be informed on which channel to expect its messages. This is accomplished with a combination of the frequency subset indicator transmitted in the system information partition, and the frequency subset number stored in the receiver. (See Figure 1.9.) Each pager will be assigned a frequency subset number (FSN) between 0 and 15. This identifies a unique subset of five frequency subset indicators (FSI). Each FSI defines the unique subset of FSNs to which the message may be directed. The FSI broadcast on a paging channel indicates that messages will be transmitted for pagers with an FSN in that FSIs subset. For example, when the FSI on a channel equals 27, only messages for receivers with FSNs of 12, 13, 14 or 15 will be carried. Conversely a receiver with FSN equals 12 should look for its messages only on channels broadcasting FSI values of 12, 22, 27, 29, or 30. The FSNs do not correspond to frequency channel numbers.

*1.10.2.3 Transmission*

Several hundred 100/250W transmitter base stations per country will initially be employed in quasi-synchronous operation. The modulation of adjacent base stations must be synchronised such that no more than 50ms differential delay is measured at the receiver. Zoning is employed and the 12 seconds, 16 batch sub-sequences (Figure 1.9) are allocated to a zone.

## 1.11 Indirect satellite paging

The SkyTel service in the US uses a geostationary satellite, West Star IV, to communicate with a satellite up-link station operating in C-band using spread spectrum modulation in California, and satellite down-link stations across the US. The caller uses the PSTN to make a toll free call from anywhere in the US to the central paging computer in Washington D.C. which then passes its calls to the up-link station in California. Selected paging calls are then transmitted via the normal local zone transmitters. The total transmission time is approximately 20 seconds and it is claimed to have 85% coverage of the US population. Redundancy is built into the system

Radio paging 43

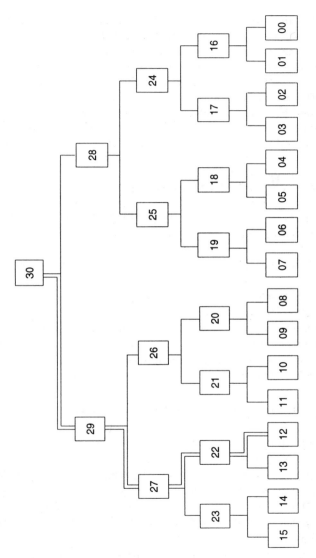

**Figure 1.9** ERMES frequency subset indicators

by having a second satellite which operates in the Ku-band. The bit error rate of the system is approximately one error in ten million bits sent. The system can be accessed from the UK by telephoning the Washington number and the u.h.f. pager can be hired in the UK or the US.

The ground transmitter frequency used is 931MHz and this frequency has also been allocated or reserved in Canada, Singapore, Malaysia, Shanghai, Bolivia, Brazil, Uruguay, Ecuador, Peru, Sri Lanka and Venezuela.

## 1.12 References

Alho, K. (1995) ERMES — roaming becomes reality, *Mobile Europe*, March.

BT (1980) *A standard code for radio paging*, report of the Post Office Code Standardisation Advisory Group (POCSAG), June 1978 and November 1980.

BT (1986) *The book of the CCIR Radio Paging Code No. 1*, Radiopaging Code Standards Group.

Communicate (1995) Caller pays widens paging's appeal, *Communicate*, March.

French, R.C. (1984) A high technology v.h.f. radio paging receiver, *IEE 1984 Conference on Mobile Radio Systems*, September.

IEC (1987) No. 489, Part 6, 2nd Ed.

Komura, M. et al. (1977) New radio paging systems, *Japan Telecom Review*, **19**, July.

Lundstorm, H. (1994) The potential for ERMES, *Mobile and Cellular*, September.

Lynch, A.E. (1993) The international growth of paging, *Global Communications*, May/June.

Makitalo, O. and Fremin, G. (1970) New system for radio paging over the FM broadcasting network, *TELE (English Ed.)* **XXII**, (2).

MTN (1994) The UK paging networks, *MTN*, 1 April.

Odneal, M. (1994) Satellite control links for paging systems, *Cellular & Mobile International*, July/August.

Okonski, R. (1994) The paging alternative, *Mobile and Cellular*, May.

Okumura, Y. et al. (1968) Field strength and its variability in v.h.f. and u.h.f. land mobile radio service, *Review of the Electrical Communications Laboratory*, **16**, (9 and 10), September/October.

Sandvos, J.L. (1982) A comparison of binary paging codes, *IEEE Vehic. Tech. Conference*, May.

Tassan, L. (1995) More power to paging elbow, *Mobile Europe*, March.

Tridgell, R.H. (1982) The CCIR radio paging code no. 1, a new world standard, *IEEE Vehic. Tech. Soc. Conference*, May.

Tridgell, R.H. and Denman, D. (1984) Experience of CCIR radiopaging code no. 1 (POCSAG) for message display paging, *IEE 1984 Conference on Mobile Radio Systems*, September.

# 2. PMR and trunked radio systems

## 2.1 Private Mobile Radio

Private Mobile radio, or PMR, has developed substantially over the years, from simplex AM communication on a single channel employing crystal controlled equipment, to that of today's multiple channel systems employing synthesised transceivers operating under microprocessor control, often with intelligent channel allocation from off-air received signalling (Baterson, 1993; Mendoza, 1994; Mulford, 1994; Webb and Shenton, 1994).

### 2.1.1 Spectrum usage

The majority of PMR services employ sub-bands within the frequency range 30MHz to 960MHz, these being periodically agreed internationally in the World Administration Radio Conference. It must be noted however that the use of these bands are not common throughout the world, as the radio regulatory bodies of individual countries often administer the use of the v.h.f. and u.h.f. spectrum to their own needs, based upon a 'non-interference' operation with other countries. A typical case is that of the Band III range of 174MHz to 225MHz, which is used in the U.K. for trunked PMR but in the neighbouring country of France for Television Broadcast services.

### 2.1.2 Equipment

Spectrum conservation techniques in mobile radio equipment used under digital agile frequency control combined with digital signalling have thus made standard the use of LSI control circuitry, with resultant sophisticated two way communications equipment becoming relatively smaller each year.

## PMR and trunked radio systems 47

### 2.1.2.1 *Frequency generation and signal processing*

Single IC frequency synthesisers with internal u.h.f. prescalers are commonly used, whereas previously these were comprised of a separate ECL prescalar, programmable divider, reference divider, and phase comparator/loop filter. The use of a relatively high first IF (Intermediate Frequency) with advanced filtering methods allows the use of lesser filtering stages in the r.f. front end stages to obtain the required band selectivity, thus minimising the physical size of the receiver circuits. Digital filtering at the second IF, typically 455kHz, may be used to allow the selection of IF selectivity under software control, to allow interchangeability of equipment between 25kHz, 20kHz, 15kHz and 12.5kHz r.f. channel spacings. Digital control of transmitter processing and resultant frequency deviation may be performed using a single IC as opposed to manual potentiometer adjustments.

### 2.1.2.2 *Tuned circuits*

Varicap controlled tuned filters under software control may be used in combination with D/A converters in the r.f. stages, to allow automatic factory alignment to be performed by digital means. The parameters for each channel or frequency range are stored in the transceiver's memory circuitry and are retrieved according to the channel selected. This may be coupled with active tuning of transmit stages to provide efficient operation with minimal generation of spurious signals, thus further reducing the size and power consumption of transceivers.

### 2.1.2.3 *Battery requirements for portable equipment*

In the case of portable equipment, the rechargeable battery source is often the limit in the equipment's overall size and hence portability. Developments in reducing portable equipment size are likely to concentrate in this area. Nickel hydrogen cells are being investigated as an alternative to nickel cadmium, the charge/discharge voltage gradient of these being similar to that of nickel cadmium thus allowing retrospective fitment. Nickel hydrogen offers approximately twice the capacity/volume ratio and four times the capacity/weight ratio of nickel cadmium. Lithium cells are commonly used in the limited number of cases where rechargeable capability is not required, these cells also having a significantly higher capacity ratio compared with nickel cadmium cells.

## 2.2 PMR systems

### 2.2.1 Simplex

The simplest form of personal mobile communication is the use of a single 'open' frequency for both transmission and reception, this normally being termed a 'simplex' channel communication method. Here all parties to the communication use a common frequency, with all users within range having communication ability, as shown in Figure 2.1.

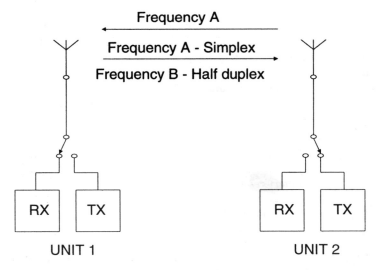

**Figure 2.1** Simplex operation

## 2.2.2 Half duplex

Progressing from this, with transmit/receive switching in use at both stations, a 'half duplex' mode occurs where personal or mobile units communicate using a different transmission frequency to that used for reception. The central radio base station uses the reverse pair of these frequencies to thus enable communication between itself and all mobile outstations. Although at first this may seem inefficient use of the spectrum, by suitable frequency management linked with geographical base station planning, more efficient use in fact may be made due to planned frequency re-use in adjacent areas.

## 2.2.3 Full duplex

Where simultaneous reception and transmission is required, this may be achieved in split frequency duplex mode by suitable filter circuits comprised within or external to the radio units, as shown in Figure 2.2. This increases the physical size of the equipment somewhat due

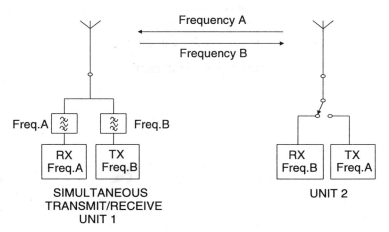

**Figure 2.2** Full duplex operation employs simultaneous reception and transmission by the use of suitable r.f. filtering

to the high-Q resonant filters required, the size of these normally being in proportion to the isolation required and in an inverse proportion to the frequency range used (DTI, 1987).

It is common practice in personal radio communication usage for the base station unit to operate in full duplex mode (DTI, 1981a), with mobile outstations operating in half duplex mode. This method allows the base to control communication, by disallowing uncoordinated portable to portable communication and allowing users with emergency messages to interrupt the communication. By coupling the received audio from the base station receiver to the base station transmitter modulator, 'talkthrough' operation may be achieved. Here the base station transmitter with its normally advantageous site over the portable/mobile stations is used to significantly extend the range of communication to a defined coverage area. Figure 2.3 shows this concept, where receive audio is linked to the transmitter for re-transmission under manual or automatic signalling control. Simplified

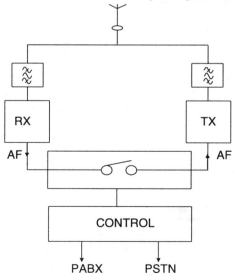

**Figure 2.3** Talkthrough operation. Receive audio is linked to the transmitter for re-transmission under manual or automatic signalling control

PABX and PSTN telephone interconnect is also feasible with this system, with the portable unit controlling dialling and call clear down through analogue or digital signalling.

### 2.2.4 Remote base station

To achieve a controlled radio coverage range in the v.h.f./u.h.f. spectrum used, a defined aerial site is normally required which may be significantly at variance with the physical location of the base station operator's console. In the case of an exclusive or shared channel allocation, a leased landline may be used to control a remotely sited base station transmitter/receiver, alternatively a point to point u.h.f. or s.h.f. radio link may be used where geographical and propagation limitations allow.

### 2.2.5 Communal base stations

A progression from the remotely sited base is a 'Communal Base Station' (DTI, 1988a). This operates in an automatic talkthrough mode in full duplex on a shared user basis. Several groups of users, each having inter-communication within their own fleet, employ the facilities of the remotely mounted equipment in a time shared mode. To ensure privacy between fleets, analogue or digital selective signalling is employed. This mode of operation is based upon the radio usage of each fleet occupying only a small percentage of available air time on the channel.

To prevent waiting users from other fleets being denied access for unreasonably long time periods, an electronically controlled time out timer is normally employed. Circuitry controlling this is fitted both to the mobile transmitter circuitry to limit the maximum length of user transmission, and to the remote base station to limit the length of air time for a given user fleet in the time slot provided.

### 2.2.6 Quasi-synchronous operation

An area or nationwide system using quasi-synchronous operation may be employed using cellular coverage on a single given channel, often combined with base station receivers having voting circuitry fitted to continuously detect and transmit received signal strength indications to the central control switch. Each transmitter is operated at a small but accurately controlled frequency offset from that of its neighbour, to combat the effects of phase cancellation in received signals at mobile unit locations (Philips, 1980). This frequency offset is typically 5Hz to 6Hz in a v.h.f. system operating with 12.5kHz channel separation, and 3Hz on u.h.f. systems. In areas where two base station transmitter signals overlap, a beat note of this frequency will be present if both signals are within the capture effect range of the mobile receiver. The probability of a complete null occurring diminishes where three or more signals overlap. Figure 2.4 gives an example of a typical coverage system.

Using these techniques, a wide coverage system based on cellular coverage techniques may be achieved without the need for a relatively advanced central processing switch, nor the use of frequency agile mobile or terminal units. Channel re-use may be achieved through the

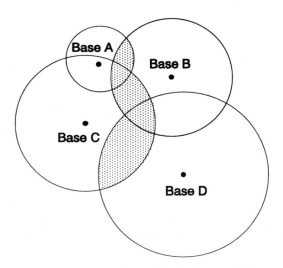

**Figure 2.4** Quasi-synchronous operation employs stable transmitters operating at a small frequency offset. Signal overlap produces received signal phase additions and cancellations

use of selective base transmitter operation dependent upon voting signal levels received from individual base station receivers.

## 2.3 Trunked mobile radio

### 2.3.1 Spectrum re-use

Over the years several improvements in spectrum conservation in PMR bands have been made in an effort to use available frequencies to the maximum benefit possible. In many developed countries, severe overcrowding of available PMR channels is encountered, especially in urban areas, and progress is continually being made towards better channel utilisation. This overcrowding has frequently led to the use of shared radio channels in a given area, with selective signalling methods used to allow a degree of communication to occur between users without an unacceptable lack of privacy.

In some cities, e.g. London in the U.K., the entire v.h.f. private mobile radio allocation has been already filled to capacity with no further channels being available for new users, and the U.K. System 4 v.h.f. radiophone system some time ago was terminated due to overcrowding. Other frequency bands such as u.h.f. as used for private mobile radio communication also suffer from congestion to varying degrees, and in some countries further u.h.f. allocations such as 900MHz are used to supplement existing 450MHz allocations for PMR use.

### 2.3.2 Trunked systems

A trunking system typically used for mobile communication employs similar basic principles to landline telephony trunking, where several radio channels are used to allow more efficient use of the available spectrum by sharing a 'pool' of channels with other users. This is the basis for most current analogue radio trunking systems in use. The application of Erlang C may be used (Erlang, 1918) to provide approximate performance figures and hence grade of service (DTI,

1986), as in Equation 2.1, where $P_d$ is probability of delay, C are the number of trunked channels, and A is the traffic load in Erlangs.

$$P_d = \frac{A^C}{A^C + C!\left(1 - \frac{A}{C}\right) \sum_{k=0}^{k=C-1} \frac{A^k}{k!}} \qquad (2.1)$$

Table 2.1 gives indications for typical 5, 10, and 20 channel mobile systems. In typical usage on a large system a dedicated 'control'

**Table 2.1** Mobile trunked system preformance

| No. of Channels | Grade of Service | Traffic (Erlang) | | No. of Mobiles | | Mean waiting time(s) |
|---|---|---|---|---|---|---|
| | | Per channel | Total | Per channel | Total | |
| | % | a | A | | | |
| 5 | 5 | 0.645 | 3.22 | 116 | 580 | 3.3 |
| | 10 | 0.719 | 3.59 | 129 | 645 | 5.8 |
| | 30 | 0.846 | 4.23 | 152 | 760 | 16.8 |
| 10 | 5 | 0.793 | 7.93 | 143 | 1430 | 3.8 |
| | 10 | 0.839 | 8.39 | 151 | 1510 | 6.2 |
| | 30 | 0.914 | 9.14 | 165 | 1645 | 16.5 |
| 15 | 5 | 0.853 | 12.79 | 153 | 2300 | 4.1 |
| | 10 | 0.886 | 13.29 | 159 | 2390 | 6.5 |
| | 30 | 0.939 | 14.09 | 169 | 2535 | 16.4 |
| 20 | 5 | 0.885 | 17.70 | 159 | 3185 | 4.3 |
| | 10 | 0.911 | 18.22 | 164 | 3280 | 6.7 |
| | 30 | 0.953 | 19.06 | 172 | 3430 | 16.4 |

channel is permanently enabled, this carries signalling information, the protocol being dependent upon the system employed, for reception by frequency scanning mobile units. In small usage systems the control channel may also be used as a traffic channel when required. This improves channel utilisation at the expense of losing instantaneous call processing signalling during peak traffic periods.

### 2.3.3 Trunked area coverage

Trunking may usefully be combined with area coverage techniques for use with frequency agile radio units under synthesiser control. Here the required area coverage is sectored, with adjacent sectors being allocated different frequencies rather than the previously described quasi-synchronous coverage method operating on a single frequency.

From carefully tailored planning of individual cell coverage patterns, frequency re-use may be employed due to capture effect performance of mobile receivers together with sufficient geographical separation of cells operating on the same frequency. Due to the tailored coverage required, either v.h.f. for wide area coverage, or u.h.f. for more closely defined coverage, is utilised.

### 2.3.4 Cell clusters

A regular polygon shape for a given cell coverage is convenient for use when planning u.h.f. trunked systems, with clusters of adjacent cells thus easily modelled for frequency re-use strategies. It must be borne in mind that local geography may radically alter this coverage shape in practical cases, however the commonly used hexagon will be used for the purposes of demonstration.

This hexagon will allow a wide degree of cluster sizes which fit together in a repeating pattern. The integer number of cells in each cluster may be given by Equation 2.2, where a and b are any positive integers and $a \geq b$ (Parsons and Gardiner, 1989).

$$C = a^2 + ab + b^2 \qquad (2.2)$$

This yields typically used cell clusters of 3, 7, and 12 cells as shown. Where a given number of channels are available for use in a dedicated trunked system, these may be divided for use in repeating numbers of cell clusters, the number of cells within each cluster depending upon the number of channels available together with acceptable degradation requirements between each cluster. This strategy is also dependent upon signal propagation, communication bandwidth, and signalling methods used. Different numbers of cells per cluster may be combined also in a typical wide area system where urban coverage is required together with rural coverage, each cell thus being tailored to the required parameters.

### 2.3.5 Base station coverage

Although mobile units often effectively have an omnidirectional aerial radiation pattern, efficiencies may be achieved in the overall number of base station sites required by the use of directional aerials. V.h.f systems are often used with omnidirectional or slightly tailored radiation patterns where local topography dictates to provide wide area coverage with a minimum number of base station sites, whilst those operating on u.h.f. (typically 450MHz and 900MHz systems) commonly use controlled aerial radiation patterns to give multiple cell clusters from a single base station site.

On elevated aerial sites in urban environments a degree of beam down-tilt is normally required on u.h.f. for effective coverage at ground level. This practice may also be usefully employed in achieving maximum frequency re-use with closely defined coverage in a given urban area supplementing that in rural and semi-rural areas.

### 2.3.6 Mobile roaming

Mobile units may roam throughout the area covered by the network of the trunked radio system base station sites, the mobile receivers being commanded to hunt for a control channel with sufficient signal strength in the area of operation. An RSSI (Radio Signal Strength Indicator) circuit in the mobile unit constantly monitors the strength of the permanently enabled control channel base transmitter, alternatively or also a bit error check is performed on received data, to establish a usable control channel.

An optional variation in a regional cellular trunked system for more efficient utilisation of a limited number of channels for combined control and traffic utilisation is that of a 'time shared' control channel. Here a dedicated signalling channel is time shared amongst a number of control transmitters operated in a cyclic manner, mobiles thus respond to the individual base transmitter giving adequate reception strength for their area.

Semi-duplex or full duplex operation is used, the base transmitter control channel being termed the Forward Control Channel (FOCC) and the corresponding mobile transmit channel being termed the Reverse Control Channel (RECC).

### 2.3.7 Control channel recognition

Once a signalling channel has been identified, the signalling information is examined by the mobile and a check performed to ensure the correct system, control category, and area or zone information is present, dependent upon the mobile unit's individual network personalisation.

If the received data is valid, the mobile locks on to this channel and, dependent upon the system being used, sends a registration identification to the relevant base station. This updates the system records on the presence and area of operation of the mobile unit. If the unit is not registered onto the system being received, the mobile relinquishes that channel and commences a further control signalling channel search.

### 2.3.8 Handoff usage

On moving away from the recognised base control transmitter, a low RSSI indication or a high bit error rate in the mobile unit causes the mobile to hunt for other acceptable control channels (Bye, 1989). The roaming mobile unit thus re-registers as required onto different base transmitter areas when moving from area to area. The system network switch updates its records to store the traffic area and channel the mobile is currently active on, hence routing incoming call indications to the mobile on the appropriate site and channel being used.

### 2.3.9 Communication

Once a call set up has been made, further signalling continues either on the control channel until the called party answers, alternatively a traffic channel is allocated for further call processing dependent upon the system being used. During communication, speech combined with supervisory signalling is employed.

Where a mobile is travelling from one cell coverage area to another, a 'handoff' action may occur where the base station signals the mobile to move its operational channel to that used by the neighbouring cell.

Here the base station in use polls adjacent cell sites to examine the signal strength being received from the mobile, and if traffic channels are available on the relevant cell site the mobile is instructed to handoff accordingly. During the communication, periodic 'maintenance messages' may also be processed to ensure correct channel utilisation and communication takes place.

# 2.4 Band III

## 2.4.1 Interface specification

Commonly called 'Band III' due to its intrinsic use in the U.K. trunked PMR system operating in the 174MHz to 225MHz Band III (HMSO, 1984) range of the spectrum, the MPT1327 signalling specification (DTI, 1988b) for trunked mobile use chosen as the mandatory protocol is also currently in use by several countries operating on Band III and other v.h.f./u.h.f. bands.

Although subject to intellectual property rights, an agreement has been achieved for the specification's use by CEPT manufacturers together with its use by system operators (Pascoe, 1989). The MPT1327 protocol is currently used in the 165MHz, 200MHz, 450MHz and 900MHz bands.

## 2.4.2 Systems

Together with public access national systems currently in use (Oliver, 1989), these based upon cellular coverage techniques combined with FDMA trunking use, public regional systems are in operation covering major cities. Large private users, e.g. airport authorities (Armstrong, 1989), employ this protocol for efficient usage of a limited number of radio channels on dedicated networks which may also use multi-site operation for area coverage requirements.

The system allows:

1. Wide area coverage with handoff capability.
2. Individual calls.
3. Fleet and sub-fleet calls.
4. Inter-fleet calls.
5. PABX interconnection
6. PSTN interconnection.
7. Off-air call set up.
8. Automatic call queuing.
9. Off-air call progress tones from digital signalling received.
10. Emergency and priority call modification.
11. Status messages.
12. Short and long data messages.
13. Call diversion.
14. Queuing of calls for automatic callback.
15. Indication of calling party identification.
16. Receiver co-channel interference protection.

## 2.4.3 Half/full duplex

Base station control channels operate either on a dedicated channel basis, on a rotating channel basis with 'Move' commands instructing mobiles to change channel accordingly, on a time shared basis where a channel is cycled between multiple transmitter sites, or with a combination of these dependent upon system requirements. Base stations for both control and traffic channel purposes operate on a full duplex basis, mobile units operate using half duplex with 1200bit/s FFSK signalling and FFSK data communication.

### 2.4.4 Control channel

A Control Channel System Codeword (CCSC) is continuously transmitted, giving details of the system's network, area, zone, control category etc. Mobile units scan the available channels stored in a programmed 'preferential hunt list', followed by an optional 'comprehensive hunt', until the correct control channel matching the unit's personalisation is confirmed.

The CCSC operates on a 'Slotted Aloha' basis, with Aloha messages inviting mobile units to register and/or send message requests, indicating the number of available message 'slots' available. Mobiles transmit randomly timed messages using FFSK to coincide with the available slots, in an effort to overcome data collisions. Upon receipt, the base station sends an acknowledgement to the calling mobile thus confirming correct reception.

A queuing system is used where the network operates on a 'first in, first out' basis together with call time limitation thus ensuring short waiting times for communication using a traffic channel. Dynamic timing is also possible using control channel 'Broadcast' messages which internally set mobile unit call timers dependent upon the traffic density at any particular time. If a user completes his allocated maximum communication period before reaching the end of the desired message, further traffic channel time is possible by entering the queue again.

Appendix A2.1 gives a selection of available MPT1327 messages thus indicating the flexibility of the network compared with alternative systems, Figure 2.5 showing the address codeword structure.

### 2.4.5 Call processing

Once a successful RQS has been made, the called party is polled with an AHY message on the control channel the called mobile is active on at any time. An ACK from the mobile indicates it will accept calls, otherwise alternative messages, such as ACKB indicating the call will

PMR and trunked radio systems 61

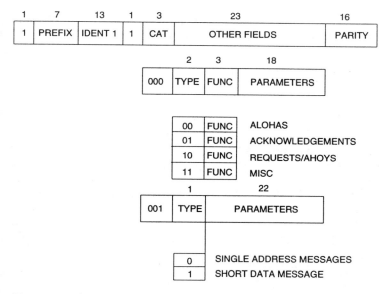

**Figure 2.5** Sample MPT1327 trunking system addresses codeword structure

be queued in the mobile memory for callback, are sent for information to the calling party. The called party is then either alerted for subscriber availability status, or queued for traffic channel availability, both parties receiving a GTC message when a traffic channel at each site is available. PABX originated calls may be made with a dedicated link between the user PABX and the Trunking System Controller (TSC), likewise with PSTN calls.

Once the mobile is present on the traffic channel, periodic FFSK maintenance messages from both the mobile and base may be processed, indicating user pressed 'on', 'off', and mobile identifications. An AHYC message requests the mobile to transmit an FFSK data message containing the unit identity and its individual Electronic Security Number (ESN), a non-registered or incorrect user receiving a CLEAR message from the TSC. Throughout communication on the channel, periodic identification fields from the base may be sent, an incorrect field received causing the mobile to clear. This ensures the required privacy from other users during co-channel reception from other cells' areas.

### 2.4.6 Data messaging on Band III

Status messages (SDM1) and short data messages (SDM2) employ use of the control channel only, the ACK protocol combined with automatic data re-transmission providing integrity of transmitted data reception. SDM1 may be linked for automatic status transmission of vehicle activity for fleet communication (GEC, 1990), and/or rapid digital messaging for fleet communication. 32 RQQ status codes are provided for, with RQQ (00000) and RQQ (11111) used for 'off hook' and 'on hook' status signalling.

SDM2 may be used for dispatcher originated short data messages entered via a terminal, such as delivery and order details. Mobile usage of SDM2 typically employs a bar code reader for work progress and parts ordering/updating.

Traffic channel data transfer is possible following either manual or automatic call set up, an error correcting protocol such as an X.25 derivative is commonly used to prevent data corruption from periodic maintenance messages. Otherwise, no defined protocols are required as an open channel is available as in the case of analogue cellular, thus base computer interrogation may be performed with added security through data encryption techniques.

## 2.5 The use of troposcatter

### 2.5.1 Introduction

With the use of high transmission power combined with high gain aerials, it is possible for v.h.f./u.h.f. signals to propagate beyond the normal 'line of sight' range. This method makes use of the availability of random irregularities in the refractive index structure of the tropospheric layer, where the refractive index differs from that of the surrounding area.

PMR and trunked radio systems 63

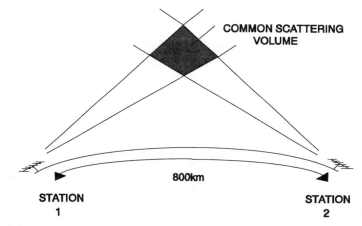

**Figure 2.6** Tropospheric scatter

A 'common scattering volume' as shown in Figure 2.6, is used where a faint signal illumination occurs at the reception end of the path. The title of 'troposcatter' is also more commonly known in commercial fields as 'forward scatter'.

### 2.5.2 Equipment requirements

The propagation medium involving the forward scattering volume involves a large transmission loss, and it becomes necessary to use high gain, narrow beamwidth aerials for both transmission and reception. The effect of the scatter angle between the receiving and transmitting beam is significant and is kept as small as possible by choosing transmitting and receiving sites so as to have an unobstructed view of the horizon.

As such, tropospheric scatter is normally limited to point to point links for medium to long distance communication. A common use is that over a sea path between a land based station on a cliff top and an offshore platform to provide telephony and other communications services.

## 2.5.3 Signal levels

The troposcatter signal received at the distant fixed point may continuously vary in signal level due to the presence of the randomly varying parameters involved in the scatter process. Variations in signal level may reach 20dB with monthly, daily, and hourly variations often being encountered (Orr, 1987).

Providing sufficient effective radiated power output exists from the transmitter, the maximum attainable communication range is in the order of 800km. This is limited by the maximum height of the useful common scattering volume in the troposphere. Although no critical frequency is involved with the scattering mechanism, the intensity of the scattered reflections decreases with increasing frequency, thus aerial gain and transmitter power must be correspondingly increased with increase of frequency.

## 2.5.4 Tropospheric ducting

Tropospheric ducting of v.h.f./u.h.f. signals leads to extended communication range over a given geographical path. Ducting is the result of the variation of the refractive index of the atmosphere found at the periphery of air masses of differing temperatures. A temperature inversion promoting tropospheric ducting occurs when a mass of cold air is overrun by a mass of warm air, and this periphery may extend in excess of 1000km to 2000km along a stationary weather front.

Temperature inversions frequently occur along coastal areas bordering large bodies of water, due to the natural onshore movement of cool, humid air shortly after sunset when the ground air cools more quickly than upper air layers. The same action may take place in the morning when the rising sun heats the upper air layers. Ducting over water has produced v.h.f. communication in excess of 4500km.

## 2.5.5 Limitations to PMR and trunked systems

Unlike troposcatter, tropospheric ducting cannot normally be controlled, but its presence should be borne in mind when planning PMR services. Disruption to local community repeater and trunking services may occur due to strong received signals on the same channel frequency from users in other areas, and along the south and east coasts of the U.K. this phenomenon is often experienced by PMR users on both v.h.f. and u.h.f. Careful frequency planning together with suitable digital or analogue selective signalling techniques may be used to lessen disruption to communication services.

## 2.6 Transmission standards

Under the Telecommunications Convention, classes of emission are designated by groups of a minimum of three characters. Currently the majority of PMR services on v.h.f./u.h.f. employ narrowband FM telephony (F3E) as the principal modulation method. AM telephony (A3E) is used in some cases, often to retain compatibility with existing systems but also for emergency services where it is advantageous for the base station control operator to be aware of a weaker mobile station attempting to establish communication whilst the channel is occupied by a stronger signal.

Single sideband telephony (J3E) is used in some cases in an effort to attain greater usage of available frequency spectrum, with some systems employing pilot carrier SSB to aid frequency synchronisation between the transmitter and remote receivers. These modes of communication are extensively documented elsewhere.

A variety of signalling methods are used in addition to the speech modulation to provide control of the communication.

## 2.7 Analogue signalling

### 2.7.1 DTMF

Dual Tone Medium Frequency signalling, which is comprehensively documented elsewhere in this handbook, is commonly used for simple numerical signalling from the mobile unit in commanding the base either to route calls to a required PABX or PSTN number, or in a more limited number of cases to selectively address other mobile

## 2.7.2 CTCSS

The Continuous Tone Controlled Squelch System (DTI, 1978) is a basic form of selective signalling which is extensively used throughout the private mobile radio field (Figure 2.7). In typical usage a single 'sub-audible' tone frequency is transmitted at 10% to 15% of the maximum system FM deviation, along with the user's speech at 100% system FM deviation, to control a tone decoder at the distant receiver. The audio output of the receiver is bandwidth filtered to typically pass frequencies above 300Hz only, thus suppressing the received tone from the user.

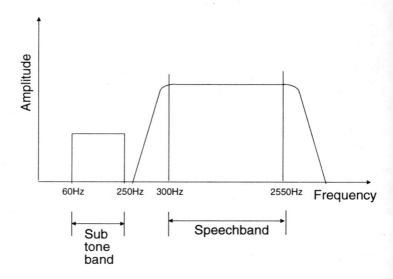

**Figure 2.7** Continuous tone controlled signalling system employs tone frequencies below 300Hz

In normal personal communication use this tone is used in addition to the carrier or noise squelch system inherent in personal receivers to control the loudspeaker muting circuitry. Thus if a signal is received with either no tone or an incorrect tone, the user's receiver remains muted to allow privacy between users of a shared radio channel in a given geographical area.

For telephone interconnection or communal base station operation, through a given dedicated or common base station, this tone is used to control line/talkthrough seizure together with user authorisation, billing and time limitation switching. Mobile units may optionally have circuitry present to also inhibit transmission whilst receiving an incorrect tone, thus preventing interference to other users of a shared channel. An internationally used set of sub-tones are given in Table 2.2.

**Table 2.2** CTCSS frequencies

| | | |
|---|---|---|
| 67.0  | 107.2 | 167.9 |
| 71.9  | 110.9 | 173.8 |
| 74.4  | 114.8 | 179.9 |
| 77.0  | 118.8 | 186.2 |
| 79.7  | 123.0 | 192.8 |
| 82.5  | 127.3 | 203.5 |
| 85.4  | 131.8 | 210.7 |
| 88.5  | 136.5 | 218.1 |
| 91.5  | 141.3 | 225.7 |
| 94.8  | 146.2 | 233.6 |
| 97.4  | 151.4 | 241.8 |
| 100.0 | 156.7 | 250.3 |
| 103.5 | 162.2 | |

## 2.7.3 Sequential tone

To overcome the effects of Rayleigh fading in a mobile environment, sequential tone signalling may be used (DTI, 1981b). Here a unique single tone corresponding to each digit is sequentially transmitted, together with a unique 'repeat' tone in place of the digit tone in the case of repeated digits. Strings of repeated digits are thus sent in the manner digit, repeat, digit, repeat etc. as shown in Figure 2.8.

The tones are sent with a defined signalling interval, this varying between 40mS and 100ms dependent upon the system being used. Thus if a tone is missed due to signal fade, the sequence is recognised as invalid and is thus re-transmitted by the mobile unit rather than being received incorrectly.

A 'transpond' system is commonly used where inter-unit dialling and status indication is performed. Here the base interrogates the mobile user, this unit acknowledges by automatically transmitting a pre-set status number to indicate availability etc. This operation may be performed automatically on a periodic basis under computer control, with the base console display being updated as required.

Alternatively the mobile unit may manually send a status update in a short time slot by entering relevant digits on the personal unit fascia, one such code normally being a request for a speech channel or

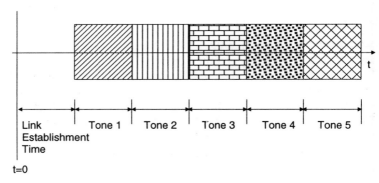

**Figure 2.8** Sequential tone signalling uses a tone sequence at commencement of transmission. A link establishment time period is required for simplex half duplex sytems

**Table 2.3** Sequential tone signalling frequencies

| Standard | EEA | ZVEI | DZVEI | ITU-R | ITU-T |
|---|---|---|---|---|---|
| Digit 1 | 1124 | 1060 | 970 | 1124 | 697 |
| Digit 2 | 1197 | 1160 | 1060 | 1197 | 770 |
| Digit 3 | 1275 | 1270 | 1160 | 1275 | 852 |
| Digit 4 | 1358 | 1400 | 1270 | 1358 | 941 |
| Digit 5 | 1446 | 1530 | 1400 | 1446 | 1209 |
| Digit 6 | 1540 | 1670 | 1530 | 1540 | 1335 |
| Digit 7 | 1647 | 1830 | 1670 | 1640 | 1477 |
| Digit 8 | 1747 | 2000 | 1830 | 1747 | 1633 |
| Digit 9 | 1860 | 2200 | 2000 | 1860 | 1800 |
| Digit 0 | 1981 | 2400 | 2200 | 1981 | 400 |
| Repeat | 2110 | 2600 | 2400 | 2110 | 2300 |
| Tone length (ms) | 40 | 70 | 70 | 100 | 100 |

telephone interconnection followed by the required number being sent by sequential tone signalling.

Table 2.3 tabulates internationally agreed tones from the EEA (Electronic Engineering Association), ITU-R (formerly CCIR), ZVEI (Zurerein der Electronissches Industrie) and ITU-R (formerly CCITT). To conform with 12.5kHz channel spacing where audio transmission is tailored to a 2.55kHz maximum frequency, a modified form of ZVEI, named 'Depressed' or DZVEI, is used.

## 2.7.4 FFSK digital selective signalling

Fast Frequency Shift Keying is now commonly used on v.h.f. and u.h.f. systems to achieve data addressing and unit identification. A

70 Data transmission

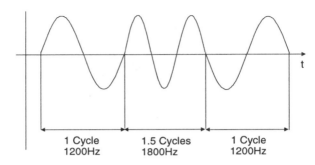

**Figure 2.9** Fast frequency shift keying employs phase continuity between digits

modulation rate generally used is 1200 bits/sec, with a binary '1' corresponding to 1 cycle of 1200Hz, a binary '0' corresponding to 1.5 cycles of 1800Hz, with NRZ coding, as illustrated in Figure 2.9. The first bit of each message commences with either phase 0 degree or 180 degrees thus ensuring phase continuity (DTI, 1981c).

Systems are generally capable of providing a data stream as shown, with a minimum transmission length of 96 bits. Transmissions commence with a preamble of at least 16 bit reversals such that the receiver demodulator can acquire bit synchronisation. This preamble is normally sufficient for transmissions where the r.f. carrier is already present, although the preamble may be continued as required to allow link establishment time in the case of simplex or half duplex operations. The following code word carries the user's identity together with address data and parity check bits to ensure correct reception, the message being re-transmitted as required in case of corruption due to Rayleigh fading.

## 2.8 Data transmission

### 2.8.1 Characteristics

Digital signalling is commonly operated in land mobile radio systems, for example data messaging on Band III systems as previously detailed. The logical progression from this is the transmission of other or longer data messages over any radio system. Typical applications include status messaging, vehicle location, pre-programmed messages e.g. of 'bar code' information, progressing up to portable and mobile data terminals employing printers and visual display screens.

A format for the transmission of digital information over land mobile radio systems has been documented in the MPT1317 code of practice (DTI, 1981c), and common usage of this employs 1200 baud FFSK using phase continuous 1200Hz (binary 1) and 1800Hz (binary 0).

The data stream is in the format shown in Figure 2.10, with a minimum length of transmission of 96 bits. A preamble for remote receiver decoder bit synchronisation is used, consisting of a minimum of 16 bit reversals, i.e. 1010...10, to be sent once the transmitter has been enabled and is transmitting power.

The length of the preamble may be extended to allow for link establishment, the preamble always ending in a logic '0'. Each message then commences with a 16 bit synchronising word as shown in Figure 2.11, to enable the decoder to establish code word forming.

Messages are transmitted in 64 bit code words, each message may occupy one or more code words as determined by the message length. Each code word contains 48 information bits and 16 check bits which

| PREAMBLE | SYNC WORD | ADDRESS CODE WORD | OPTIONAL DATA CODE WORDS |
|---|---|---|---|

**Figure 2.10** MPT1317 digital code format for data over PMR

| BIT NO. | 1 | 2 | 3 | 4 | 5 | 6 | 7 | 8 | 9 | 10 | 11 | 12 | 13 | 14 | 15 | 16 |
|---|---|---|---|---|---|---|---|---|---|---|---|---|---|---|---|---|
| BIT VALUE | 1 | 1 | 0 | 0 | 0 | 1 | 0 | 0 | 1 | 1 | 0 | 1 | 0 | 1 | 1 | 1 |

**Figure 2.11** MPT1317 synchronisation frame

| BIT NO. | 1 | 2      8 | 9                    48 | 49           64 |
|---|---|---|---|---|
| NO. of BITS | 1 | 7 | 40 | 16 |
|  | '1' | USER's ID | ADDRESS AND DATA | CHECK BITS |

**Figure 2.12** MPT1317 address code word structure

are used for error detection and data code words. There are two types of code word, address code words and data code words.

### 2.8.2 Address code word

The first code word of every message is an address code word as shown in Figure 2.12. Bit 1 is always binary 1 to distinguish the word from a data code word, bits 2 to 8 inclusive specify the user's identity, bits 9 to 48 inclusive may be assigned to represent addresses and data, and bits 49 to 64 inclusive are check bits. Within the data field, the recommended practice is for bits 9 to 20 to specify the addressee's identity, bits 21 to 32 to specify the addressor's identity, and bits 33 to 48 to be used for data.

### 2.8.3 Data code word

An arbitrary number of data code words as shown in Figure 2.13, may follow an address code word, as required to accommodate the message.

Bit 1 is always binary 0 to distinguish the word from an address code word, bits 2 to 48 are used for the data, and bits 49 to 64 are used as check bits.

Several messages may be sent in one transmission as shown in Figure 2.14, i.e. concatenated messages. In this case it is not necessary to send the preamble at the start of every message.

## PMR and trunked radio systems 73

| BIT NO. | 1 | 2 | 48 | 49 | 64 |
|---|---|---|---|---|---|
| NO. of BITS | 1 | 47 | | 16 | |
| | '0' | DATA | | CHECK BITS | |

**Figure 2.13** MPT1317 data code word structure

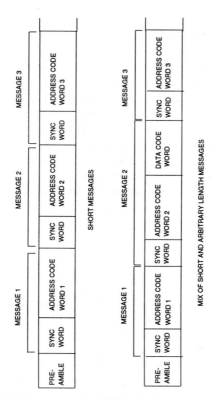

**Figure 2.14** MPT1317 concatenated messages

## 2.8.4 Encoding and error checking

The 16 check bits are calculated in three stages. Firstly, 15 check bits are appended to the 48 information bits by encoding them in a (63,48) cyclic code. For encoding, the information bits 1 to 48 may be considered to be the coefficients of a polynomial having terms from $X^{62}$ down to $X^{15}$. This polynomial is divided by modulo 2 by the generating polynomial as in Equation 2.3.

$$X^{15} + X^{14} + X^{13} + X^{11} + X^4 + X^2 + X^1 \tag{2.3}$$

The 15 check bits, code word bits 49 to 63, correspond to the coefficients of the terms from $X^{14}$ to $X^0$ in the remainder polynomial found at the completion of the division. The (63,48) cyclic code has a minimum distance of 5 and so guarantees detection of up to 4 bit errors in one code word.

Secondly, the final check bit of the (63,48) cyclic code (code word bit 63) is inverted to protect against misframing in the decoder.

Thirdly, one bit is appended to the 63 bit block to provide an even bit parity check of the whole 64 bit code word. The overall parity bit ensures that all odd numbers of errors can be detected, so that the overall 64 bit code guarantees that up to 5 bit errors can be detected.

At the receiver, each code word can be checked for errors by recalculating the check bits for the received information bits. Any differences between the received check bits and the recalculated check bits indicates that the received code word contains errors.

## 2.8.5 Packet data communication

Data communication networks employing error correcting protocols have been used non-commercially for some time now, and are currently starting to become employed for data transmission over PMR systems. Packetised data with automatic requests for re-transmission may be used to overcome effects encountered in typical PMR usage such as Rayleigh fading in a mobile environment, as well as allowing the shared use of a single r.f. channel. Interconnected packet nodes may be linked to radio base stations operating on h.f., v.h.f., u.h.f. and s.h.f. to form an international network for data store and forward operations, using both terrestrial routing and orbiting digital communications satellites for inter-continental message routing.

AX.25 (ARRL, 1984) a derivative of X.25, is commonly used together with level 2, 3, and 4 node systems.

User access employs 1200 baud FFSK operation using either Bell 202 tones or MPT1317 tones modulating an FM carrier, with operating frequencies on v.h.f. and u.h.f. TDMA is used with terminal node controllers automatically re-transmitting packets of between 32 and 255 bytes when an ACK is not received within a given time. Internode routing and satellite uplink/downlinking commonly uses 9600 baud FSK with Manchester coding (Miller, 1988) with operating frequencies in the u.h.f. regions. Coverage in many areas is such that hand portable data terminals may be successfully used, the automatic retry protocol linked to suitable packet data lengths for the intended use combating signal fading to a large degree.

## 2.8.6 Frame format

Link layer packet transmissions are sent in a number of frames, each frame subdivided into fields as in Figure 2.15. The transmission of a frame is preceded by 16 bit reversals, followed by the start flag, address field, control field, a network protocol identifier (PID), information field, frame check sequence (FCS) and an end flag.

The information field may be of a variable length, this may be set to any integral number of octets of information up to a maximum of 256. This allows the length of the transmitted packets to be controlled, thus allowing customisation to account for interpolation with other overhead messages such as the periodic maintenance messages found on MPT1327 systems. Together with a six unit alphanumeric identifier string, a seventh identifier, the SSID (Secondary Station Identifier) may be used to allow up to sixteen streams of individual data connections to be conducted from a given ID.

Commercial packet network usage is now commencing using proprietary protocols linked to data encryption for security, e.g. in the U.K. a small number of such systems are licensed for national cover-

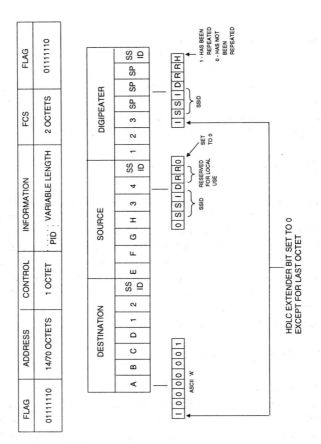

**Figure 2.15** AX25 packet frame format

age, employing nodes interlinked via landlines. Uses of similar systems for Electronic Funds Transfer are also in hand.

## 2.8.7 Satellite land mobile data communication

Low Earth orbiting satellites (Clark, 1988) allow the use of store and forward data communication to be achievable using low power transportable and mobile mounted equipment on v.h.f. and u.h.f. employing omni-direction aerials (AMSAT, 1991).

As well as one to one AX.25 usage, a 'broadcast' protocol may also be employed from the satellite where continuous bulletin information is transmitted for reception by permanently installed land terminals linked to computer controlled disk based storage, with individual repeat request messages only being transmitted by the ground station to complete 'holes' in the received packetised data.

# 2.9 Applications

Private Mobile Radio is identified thus to distinguish it from PAMR, Public Access Mobile Radio. PMR normally is used by individual fleets of users with inter-fleet communication, whereas PAMR covers public access communication such as those with PSTN access, e.g. Cellular, CT2 etc. A degree of overlap occurs however, with many PMR systems having 'private' PABX and PSTN access facilities from within their own systems, together with trunked PMR offering intra-fleet communication and PABX/PSTN access facilities, dependent upon the system in use.

## 2.9.1 Business use

PMR is thus generally used within a business environment for inter-unit communication within the business fleet, e.g. for rapid dispatcher to delivery vehicle communication. Through the use of carefully tailored base station aerial sites, communication may be usefully achieved within an area where public access systems are unable to provide a service, for example within an office or building site environment, or on a larger scale along a dedicated network of permanent ways such as the British Railways trunked PMR system now in use.

Due to the one to one communication scenario, PMR is generally considered unsuitable for private citizen use where PSTN access, either one way or two way, is often required. Here, cellular telephony, CT2, and PCN are likely to be of greater benefit. Fees levied by national cellular system operators for network calls however often make this unsuitable for users with frequent but often short calls, e.g. taxi and delivery firms operating in a given locality, where PMR normally offers a more economic solution if frequency assignments exist for this.

### 2.9.2 PMR growth

The usage of PMR is continuing to increase rather than decrease, indicating that public access systems are complimenting rather than replacing PMR systems.

The development of the TETRA standard is expected to provide a further boost to PMR systems (Harrison, 1994; Renduchintala and Razzell, 1995; Whitehead, 1995).

The extra frequency spectrum needed must thus be used to the best possible effect in congested areas, with trunked PMR becoming more prevalent.

Frequency sharing between users is under control in many countries by co-ordination work performed by national regulatory bodies, and this must be considered and complied with to ensure the continued usefulness of PMR.

# 2.10  Appendix 2.1

**MPT1327 sample command structure**

### 2.10.1  Channel allocation

GTC  - Go To Channel

### 2.10.2  Aloha invitations

ALH  - General
ALHS - Standard Data Excluded
ALHD - 'Simple' Calls excluded
ALHE - Emergency only
ALHR - Registration or Emergency

### 2.10.3 Acknowledgements

ACK - General
ACKI - Intermediate
ACKQ - Call queued
ACKX - Message rejected
ACKV - Called Unit unavailable
ACKE - Emergency
ACKB - Call-back

### 2.10.4 Radio unit requests

RQS - Simple
RQD - Data
RQT - Divert
RQE - Emergency
RQR - Registration
RQQ - Status
RQC - Short Data

### 2.10.5 TSC Ahoys

AHY - General availability check
AHYQ - Status message
AHYC - Short Data Invitation

## 2.11 References

AMSAT (1991) SatelLife — Healthnet satellite communications using transportable Earth terminals, *AMSAT-U.K. Colloquium*, University of Surrey, July.

Armstrong, R. (1995) Key issues for the mobile radio user, *Mobile Communications International*, April.

Armstrong, W. (1989) Trunked PMR in British Airways, *IBC Trunked Private Mobile Radio Conference*, November.

ARRL (1984) *Packet Radio Link Layer protocol, AX25*, December.

# 80 References

Azemard, H. (1994) TETRA — a standard for police communications, *Electrical Communication*, 2nd Quarter.

Baterson, M.J. (1993) PMR, into the digital era, *Global Communications*, September/October.

Bye, K.J. (1989) Handover criteria and control in cellular and microcellular systems, *IEE Fifth International Conference on Mobile Radio and Personal Communications*, December.

Clark, T. (1988) AMSAT's Microsat/pacsat Program, *7th ARRL Computer Networking Conference*, Columbia U.S.A., October.

DTI (1978) *Performance specification for continuous tone controlled signalling (CTCSS) for use in the land mobile services*, Department of Trade and Industry, London, MPT1306, January.

DTI (1981a) *Code of practice, requirements for duplex operation in the land mobile services*, Department of Trade and Industry, London, MPT1315, March.

DTI (1981b) *Code of practice for selective signalling for use in the private mobile radio services*, Department of Trade and Industry, London, MPT1316, January.

DTI (1981c) *Code of practice for the transmission of digital information over land mobile radio systems*, Department of Trade and Industry, London, MPT1317, April.

DTI (1986) *Engineering memorandum for trunked systems in the land mobile radio service*, Department of Trade and Industry, London, MPT1318, February.

DTI (1987) *Code of practice for radio site engineering*, Department of Trade and Industry, London, MPT1331, April.

DTI (1988a) *Code of practice for repeater operation at communal sites*, Department of Trade and Industry, London, MPT1351, June.

DTI (1988b) *A signalling standard for trunked private land mobile radio systems*, Department of Trade and Industry, London, MPT1327, January.

Erlang, A.K. (1918) Solution of some problems in the theory of probabilities of significance in automatic telephone exchanges, *Post Office Electrical Engineers Journal*, January.

GEC (1990) *Data services for fleet managers*, GEC National One.

Harrison, D. (1994) TETRA — a digitally encoded speech PMR trunking system, *Mobile Europe*, April.

HMSO (1984) *Bands I and III, A consultative document*, May.

Hobeche R.J. (1985) *Land Mobile Radio Systems*, Peter Peregrinus.

Lee, W.C. (1982) *Mobile Communications Engineering*, McGraw Hill, New York.

Mendoza, N. (1994) Trunking: a great option, *Global Communications*, July/August.

Miller, J. (1988) 9600 baud packet radio modem design, *7th Computer Networking Conference*, Columbia U.S.A., October.

Mulford, K. (1994) Signalling multiplies digital PMR features, *Cellular & Mobile International*, May/June.

Oliver, B. (1989) U.K. national network services, *IBC Trunked Private Mobile Radio Conference*, November.

Orr, W.I. (1987) *Radio Handbook*, 23rd Edition, Sama.

Parsons, J.D. and Gardiner, J.G. (1989) *Mobile Communications Systems, 1st Edition*, Blackie and Sons, Glasgow and London.

Pascoe, R. (1989) The manufacturer's viewpoint, *IBC Trunked Private Mobile Radio Conference*, November.

Philips (1980) *Quasi-synchronous operation of two or more transmitters*, Philips Radio Communications Systems Ltd. publication ref. TSP361/2, April.

Radiocom. (1990) *Mobile Radio*, U.K. Radiocommunications Agency, Annual Report 1990–1991, pp. 30–31.

Renduchintala, M. and Razzell, C. (1995) TETRA radio terminal design: technical challenges of the physical layer, *Mobile and Cellular*, June.

Rousseau, P. (1994) Digital signalling for private mobile radio (DSPMR), *Electrical Communication*, 2nd Quarter.

Shanahan, H. (1994) Understanding and testing EDACS trunked mobile radio, *Cellular & Mobile International*, May/June.

Webb, W.T. and Shenton, R.D. (1994) Pan-European railway communications: where PMR and cellular meet, *Electronics & Communication Engineering Journal*, August.

Whitehead, P. (1995) TDMA is tops for PMR, *Mobile Europe*, June.

# 3. Cordless communications

## 3.1 Introduction

Cordless telephone products first appeared as a low cost option for the home in North America during the late 1970s. Technically it was a radio extension telephone that operated at low power, to conserve operating time between battery charges, and frequencies of 1.7MHz (base to handset) and 49MHz (handset to base). The chosen technology was analogue frequency modulation which resulted in prices around $100 for a residential unit. Although range around the home was limited, typically covering an apartment or a house and gardens, the combination of low cost and freedom of use made the idea an instant success; they sold in their millions and still do.

The success was exported to Europe by way of illegal imports into the UK, France, Germany, Holland, etc. These imports infringed local telecommunications law and practice. In the UK, for example, the frequencies chosen for North American were allocated to maritime and the then broadcast television service. Furthermore their transmission characteristics gave unsatisfactory speech performance when they were connected to European public telecommunications networks. Nevertheless the quantity of illegal imports demonstrated that a market existed for such a product and consequently the European PTT's reacted by adopting appropriate standards for Europe. From this beginning the present cordless market is now on the verge of making a significant breakthrough as a natural telephone terminal apparatus in both the home, office and manufacturing environments.

### 3.1.1 Analogue technology

Two families of analogue CT have been standardised; that based on the original North American standard (h.f./v.h.f.) and a later higher capacity European development operating in the 900 MHz band.

## 3.1.2 H.f./v.h.f. analogue CT

In order to compete on equal price terms with the import of illegal North American standard product, many countries chose to adopt the same standard with or without technical design changes. In Europe, the UK and France followed this course, making changes to the design that avoided radio frequency allocation and transmission quality problems, but did not significantly increase the basic production cost.

One major change was to incorporate some form of dialling and incoming call security that ensured only the legitimate handset was able to open up and receive a call from its associated base station. This stopped the accidental, or sometimes deliberate, practice of long distance calls being dialled over a neighbour's cordless terminal with consequent misdirection of the call charge. The basic technical characteristics of the UK version of this analogue CT are as follows:

1. Eight analogue FM channel pairs operating around 1.7MHz and 47.5MHz.
2. Radiated power limited to 10mW maximum.
3. Choice of channel pair is set at manufacture but random at time and place of purchase. Line access and incoming call security by a handshake process involving at least 60000 randomly selected binary codes.

Products to this enhanced specification have been freely available in the UK since 1983. In France different frequencies were adopted (21MHz and 46MHz). More recently the North American market has concentrated both directions of transmission in the region of 49MHz in order to give more transmission channels and hence more transmission capacity. The technology, however, is basically the same and retains the very attractive low production cost.

## 3.1.3 European 900MHz analogue standard

Another body of opinion in Europe took a somewhat longer view of the potential of cordlessness, in terms of its potential demand on radio spectrum when the number of cordless terminals exceeded about 10% of normal wired terminals. This vision resulted in the following standard, that has been adopted in most CEPT affiliated countries, with the exception of the UK and France who chose the h.f/v.h.f. option that offered a significantly lower production cost. The CEPT standard has the following general characteristics:

1. Analogue FM operation.
2. Forty 25kHz duplex channels in the frequency bands 914–915MHz paired with 959–960MHz.
3. Radiated power limited to 10mW.
4. Dynamic channel selection (DCS) on call set up.

This last feature sets apart this standard from the earlier technology. On call set up either from a base (incoming call) or handset (outgoing call) all forty channels are scanned in order to select a vacant radio channel, or one where the co-channel interference is acceptable low, on which to set up the telephone call.

This feature has a complexity burden but results in all users being able to use any available radio channel. Therefore spectrum efficiency in terms of traffic carried per megahertz per square kilometre is very much increased compared to a factory set single channel system of the type considered in the last section.

The benefits of DCS shows when the total cordless traffic demand in a localised area is such that co-channel interference from other users becomes the dominating factor in terms of speech quality and range of operation.

Technically the CEPT standard is a major step forward in cordless product design but is consequently more expensive than the far simpler h.f./v.h.f. analogue design. It has, therefore, not achieved the degree of market penetration throughout Europe as its proposers had originally predicted.

## 3.2 Digital technology

The analogue cordless telephone has succeeded in opening up the market for localised mobility around the home or place of work, without incurring the costs and regulation associated with a fully mobile cellular radio service. However, its analogue nature and, in its more successful form, limited traffic carrying capabilities severely limited its full development potential.

What was needed was a ubiquitous technology that could be exploited widely in the home, office and factory, in such numbers that all could benefit from the economies of scale that mass uniform production can provide. The new generation needed the characteristics as in the following sections.

### 3.2.1 Areas of application

1. Residential, private use in houses and apartments.
2. Small CT systems for office use.
3. Large CT systems, multi-cell private switch telephony based building mobile systems with roaming and in-call handover between cells (the so-called cordless business communications system, CBCS).
4. Radio access to local public and other telecommunications networks, e.g. telepoint systems.

The CBCS opens the opportunity towards the cordless office with greater freedom to rearrange office space as work demands. This is a result of reduced telecommunications wiring needs and costs. However radio coverage over the entire service area must be good enough to ensure that the probability of a call being set up exceeds 99% otherwise users' expectations, based on the performance of wired telephones, will not be met.

Telepoint is a service provided to cordless handset owners from cordless base stations located in public places, e.g. railway stations, shopping precincts, fast-food restaurants etc.

This is a basic public communication service for the less migratory, more localised sector of the travelling market and thus does not compete directly with the wide roaming mobile cellular network. Hence a handset purchased for use in the home and/or work place can also be used to gain access to a telepoint service whilst the user is in transit between them.

## 3.2.2 Main service principles

1. The system must make provision for voice and non-voice transmission.
2. All radio channels are available to all users and applications, hence channel licensing and regulation is not needed.
3. The air interface between the handset and base has a common specification for telephony so that a common handset design can be used in all applications.
4. Speech quality equivalent to that of a wired telephone.

## 3.2.3 Traffic capacity

Research has shown (Swain, 1985) that a modest CT penetration of 7% of all telephone terminals in a city centre could produce mean traffic densities in excess of 800 Erlangs per square km. (Note an Erlang is the amount of traffic carried by one line operating for 100% of the time.) Other studies (ETSI, 1992a) indicate the following traffic densities by application:

1. Residential suburban house, 150E per sq. km at 0.05E per telephone and cordless penetration of 30%.
2. Residential apartment block, 200E per sq. km at 0.05E per telephone and cordless penetration of 30%.
3. Cordless business system, 10000 E per sq. km per floor at 0.2E per terminal and cordless penetration of 100%.
4. Telepoint service (examples of railway stations and airports), 900 – 5500E per sq. km.

## 3.2.4 Digital cordless standards

The foregoing requirements clearly point towards a new digital cordless telecommunications standard and as early as 1981 studies were in progress to identify the technical options. Since that time two digital standards have been developed within Europe to the status of interim European Telecommunications Standards by the European Telecommunications Standards Institute (ETSI), as follows:

Cordless communications 87

1. CT2/CAI; which translates to second generation CT with a Common Air Interface (ETSI, 1991).
2. DECT; which means Digital European Cordless Telecommunications (ETSI, 1992b).

The key factor in both standards, however, is that they offer a common air interface specification. This means that each can support a public access voice service in which the handset and base station may be manufactured by different companies and yet still signal and communicate to each other to establish the call. This is not a feature of the earlier analogue standards.

These two ETSI standards are considered by many to be complementary in terms of the market expected to be served and the entry time to that market. CT2/CAI has been optimised to serve the residential and small (business) cordless markets with potential to open up the high capacity business market. DECT, however, has been designed from the outset to meet the demands of the very high capacity cordless office with roaming and in-call handover as standard features of the specification. It also has the capability to be appropriately configured for the residential markets. Both have telepoint capability.

## 3.3 CT2/CAI digital specification

These specifications are described in Swain (1990), Holmes and Swain (1990) and Evans (1990). The FDMA/TDD principle of operation of CT2/CAI is shown in Figure 3.1. Duplex transmission is

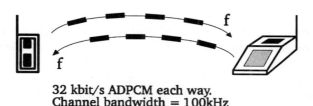

32 kbit/s ADPCM each way.
Channel bandwidth = 100kHz

**Figure 3.1** Single-carrier burst-mode duplex system

provided by transmitting in time interleaved burst mode on the same carrier frequency for both directions. This technique was appropriate for a low cost instrument since both ends are identical. Furthermore the technique only required a single block of radio spectrum rather than the duplex bands favoured by other mobile radio systems. Hence the band 864–868MHz was allocated to the service to support forty 100kHz duplex channels.

By adopting the ITU-T (formerly CCITT) standard 32kbit/s ADPCM speech coding algorithm it was possible to contain the transmitted symbol rate to 72kbit/s with some allowance for framing and signalling bits. In practice the corresponding B-channel capacity is 32kbit/s and that for the D-channel 1kbit/s or 2kbit/s depending on manufacturer's choice.

By adopting a common air interface the radio interface is standardised in terms of:

1. Physical parameters of the radio link e.g. modulation, frame rate, data rate.
2. System and user signalling.
3. Speech transmission.

### 3.3.1 Radio aspects

With only forty channels available to support the high traffic demands it was clear that the benefit of dynamic channel selection would be needed. Thus both handset and base are required to operate on any of the forty channel pairs through selection of the channel with the lowest co-channel and/or adjacent channel interference. This technique is more than adequate to meet the traffic capacity needs of residential and small business operations, but other techniques need to be exercised to raise the effective capacity to meet the requirements of larger business systems.

To increase the frequency re-use ability of the system in large buildings cordless business systems will need to use the multi-cell coverage techniques employed by cellular mobile systems. However, in a building the cells will be used on each floor, consequently the cell structure has three dimensions to it and allowance must be made for signals passing through the floors as well as horizontally through the walls of the building.

To ensure good and even coverage both antennae and radiating cables may be used to provide service in a large building complex as in Figures 3.2 and 3.3 (Holmes and Swain, 1990).

The detail parameters of the radio interface are given in ETSI (1991) but it specifies two level frequency shift keying with Gaussian filter shaping and a frequency deviation of 14.4kHz to 25.2kHz above

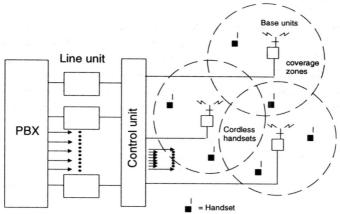

**Figure 3.2** Wireless PBX system using antennas

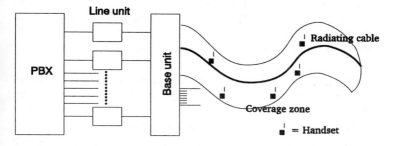

**Figure 3.3** Wireless PBX system using radiating cable

carrier frequency to represent a binary 1 and equal shifts below the carrier frequency for binary 0. This represents a modulation index range of 0.4 to 0.7. As previously noted time division duplex transmission is used with 1ms for each transmit and receive packet. A dead time between transmit and receive bursts has been created to allow transmitters to ramp up and ramp down their power in a way designed to limit spectral splatter across adjacent channels. This period also permits oscillators to settle between bursts.

### 3.3.2 Frame multiplex structure

When a call is being initiated there is no synchronisation between the handset and base, and both parts will be scanning across the forty radio channels. In order for a call to be set up and maintained CT2/CAI employs three multiplex structures during the course of a call.

Multiplex 1 is used during the normal handset to base communication phase when synchronisation and call set up have been achieved and the call is in progress. The structure is shown in Figure 3.4 (Evans, 1990) where it can be seen that two channels are supported.

The B-channel conveys the 32kbit/s speech signal (or perhaps 32kbit/s data) and the base and handset identities are conveyed through the D-channel. While D-channel errors are detected and corrected by requesting a re-transmission there is no error detection on the B-channel. The observed performance of the D-channel can be used to determine B-channel performance and if appropriate instigate a radio channel change to a better quality channel during the speech. Generally the channel change will be quick and go undetected by the user.

Figure 3.4 shows that the D-channel capacity can either be 1kbit/s or 2kbit/s according to manufacturer's choice, but this choice must be signalled between the base and handset at call set up. Clearly a public telepoint base station must be able to support both. It should be noted that the error corrected rates for each possibility is about half the basic throughput.

Multiplex 2 is used only when the base station is setting up a link with its associated handset. At this stage the B-channel is not active

## Cordless communications 91

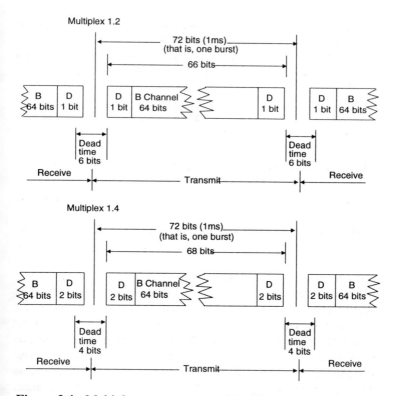

**Figure 3.4** Multiplexer structures used by CT2/CIA

but the base needs to transmit its identity to the required handset via the D-channel and synchronise that handset to its frame using the SYN-channel. The format is shown in Figure 3.5 (Evans, 1990). In this format the D-channel has a 16kbit/s capacity, the preamble consists of 1,0,1,0,1...reversals for bit timing, and the channel marker word (CHM) is used to set up frame synchronisation. Successful synchronisation is marked by the transmission of the SYNC word from the handset at which point it is able to transmit the same frame structure and forward its identification code in the D-channel.

## 92 CT2/CAI digital specification

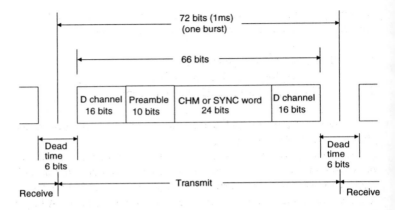

**Figure 3.5** Format used for multiplexer 2 of Figure 3.4

If the handset needs to set up a call then a different frame structure, multiplex 3, is initially used by the handset in order to synchronise with the base unit. This is important because in multiplex 1 and 2 operation the base and handset terminals alternately transmit and receive and it is necessary to avoid unsuccessful call set up due to, for example, both units transmitting at the same time. In multiplex 3 the handset repeatedly transmits a sequence of preamble bits, D-channel identity information and channel markers for a period of 10ms. This is followed by a 4ms receive period in which a response from the base unit is listened for. This process is repeated for up to 5 seconds after which call set up is abandoned. This frame structure is designed to ensure that the base station, which is operating in the 1ms transmit and 1ms receive mode, has a maximum chance of intercepting the initially un-synchronous handset transmission. After synchronisation call set up proceeds using multiplex 2.

### 3.3.3 Protocol structure

Signalling in CT2/CAI is similar in form to that adopted in Layers 1, 2 and 3 of the Open Systems Interconnection (OSI) model. Layer 1 ensures that communication channels can be established and maintained. To achieve this three channels types are needed:

1. B-channel for speech or data.
2. D-channel for signalling and control.
3. SYN-channel for bit and burst synchronisation.

At various stages of a call set up these three are conveyed by the above mentioned multiplexes.

Layer 2 supports the ability of CT2/CAI to communicate messages between end points of the link. Such messages include those associated with the D-channel error detection and correction, message acknowledgement, call set up and clear down messages etc. A Layer 2 message package consists of up to six words each formed from eight 8-bit bytes. The first word is the address code word and the remainder are data code words. Messages may be formed from a number of message packages.

Layer 3 messages are the information elements delivered error free by layer 2 to the end of the CT2/CAI radio link. These elements have meaning and can be translated into specific responses. For example keyed digits 0–9, star and square, handset display elements and many other messages that a manufacturer may wish to employ in his product.

### 3.3.4 Transmission plan and digital codec

To meet the European standard objectives it is necessary to define the transmission characteristics of both the handset and base unit. For interworking with digital networks the requirements of ETSI draft recommendation NET33 are met. In the case of analogue network interconnection, for historical reasons, the requirements change from country to country throughout Europe. So to ensure a common handset specification, particularly for telepoint application, any necessary country specific corrections to the overall speech transmission characteristic is built in to the base unit. To avoid problems arising from terminal speech echo in the transmission network the loop delay imposed on a CT2/CAI link has been set to less than 5ms.

The digital speech codec algorithm used by CT2/CAI has been taken from the ITU-T Recommendation G.721 (Blue Book) which defines a 32kbit/s adaptive differential pulse code modulation codec. Some, strictly controlled, licence is given to manufacturers to simplify the resulting codec in order to limit power requirements and complexity.

### 3.3.5 Implementation

The basic principles of CT2/CAI operation offer significant opportunities to minimise the inherent cost of the basic digital link. For example the 100kHz channelling does not require time dispersion equalisation even when operating in a multipath environment with no line-of-sight signal component. Furthermore there is no need for a radio frequency duplexer and less carrier frequency sources are required. From a speech transmission point of view the moderate processing delays involved do not require echo control circuitry or raise absolute network delay issues. With these benefits the radio circuit design is simplified and the complexity of implementation is concentrated at baseband which can made using low power consumption, large scale integration CMOS technology.

Figure 3.6 is an example of the degree of integration that is possible and which leads towards a common set of components to support all applications. This is a key objective in realising the benefits of large volume production in terms of minimised component costs.

## 3.4 DECT digital specification

Around the mid 1980s a number of companies under the auspices of the European Conference of Telecommunications Manufacturers (ECTEL) started to define and specify a digital cordless telecommunication standard for business system use. It adopted many concepts previously exploited in European CT standards but by implementing multi-carrier time division multiple access it considerably increased the potential for cordless operation particularly in a business environment. Eventually this work led to the ETSI standard

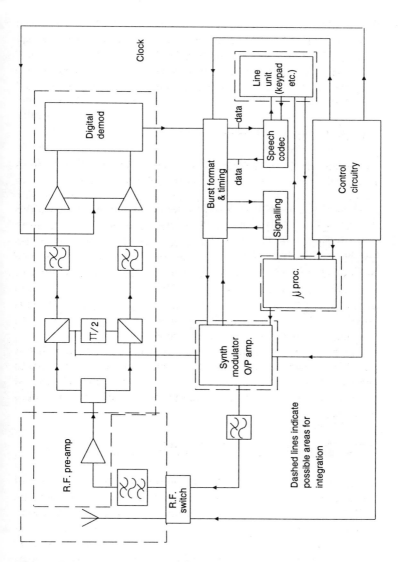

**Figure 3.6** Possible integration of CT2

## 96  DECT digital specification

that is now know as Digital European Cordless Telecommunications, or DECT for short. (ETSI, 1992b; Ochsner, 1990; ETSI, 1992c, Hendy, 1994; MTN, 1995; Koopmans, 1995.)

The DECT specification has two levels of standardisation.

1. The Common Interface (CI) specification that enables conforming equipment from different manufacturers to successfully communicate in a public access service (e.g. telepoint).
2. The coexistence interface specification that allows proprietary, non-CI, standard equipment to coexist in the common spectrum resource.

Thus DECT is able to support both public access requirements and the proprietary needs of the manufacturers, particularly those interested in exploiting the considerable potential of cordlessness in the business environment. This is shown in Figure 3.7 (ETSI, 1992b), the layers being described in the next section.

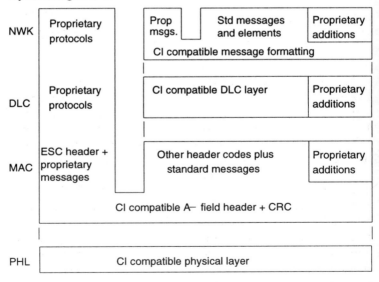

**Figure 3.7**  Structures of escape routes within the Common Interface

**Table 3.1** Data requirements in DECT

| Application | Access latency | Transaction duration at full rate | Full rate (without errors) |
|---|---|---|---|
| Remote terminal | | | |
| Text | 50ms | 100ms – 5 secs | 10 – 20kbit/s |
| Graphics | 50ms | 500ms – 10 secs | 24 – 128kbit/s |
| Batch file transfer | | | |
| Light | 1 – 5 secs | 1 – 30 secs | 32kbit/s |
| Heavy | 1 – 30 secs | 5 – 1000 secs | 64kbit/s |
| Real time file access | | | |
| Slices | 50ms | 200ms – 2 secs | 64 – 256kbit/s |
| Chunks | 500ms | 1 – 10 secs | 64 – 256kbit/s |

The application areas of DECT are somewhat wider than those identified in Section 3.3. For example the system is expected to support radio access to office voice and data networks at information bit rates considerably in excess of the 32kbit/s foreseen for speech transmission. Indeed data transmission has figured significantly in the preparation of the specification. DECT, therefore, offers bearers that are well matched to the needs of teleservices.

For ISDN-based applications a continuous 144kbit/s full duplex bearer is available. Even this is not sufficient for other uses associated with data transmission in real time and/or short bursts. Such requirements are indicated in Table 3.1 (ETSI, 1992b). For data applications, variable transaction times 100ms to 10 seconds are anticipated and transmission is anticipated to be predominantly one way. Fast link establishment time, under 50ms (not including Portable Part verification) is required. Variable rate communications is required. Note that these applications demand rapid access to bearer channel and since radio channels must be released between bursts of information (to conserve spectrum for other users) then the result is a requirement for rapid radio channel acquisition algorithms capable of seizing a channel in less than 50ms.

Although the above has been written in terms of DECT as a piece of cordless terminal equipment, it was made plain during its development that the technology can also be used as an access technology to other communication networks both public and private. For example, provision has been made in the design of DECT to enable it to be developed as another access technique able to use the GSM digital cellular fixed network. This means that DECT must in due course be able to access and use the roaming and location intelligence built in to the GSM system. Similar provisions are being made to ensure the capability to interwork with evolving public intelligent telecommunication networks. Thus DECT is not just another cordless telephone peripheral to telephone networks, it also has the capability to become an access technology that is integrated with the network.

### 3.4.1 DECT protocol architecture

The DECT protocol (ETSI, 1992b) is based on the principles of the OSI model. The complete common interface relevant to public access applications is defined in terms of the three lowest layers modified to account for the specific requirements of radio transmission and in-call handover. The structure is shown in Figure 3.8 where four DECT layers are identified. The OSI layers are indicated for reference. The Physical Layer (PHL) has the task of modulation and demodulation, acquire bit and burst synchronisation, control synchronisation and independent burst collision detection and measure the received signal strengths.

The Medium Access Control Layer (MAC) performs two main functions:

1. It selects the radio channels and then establishes and releases the communication link.
2. It multiplexes and demultiplexes all information into burst packages.

Cordless communications 99

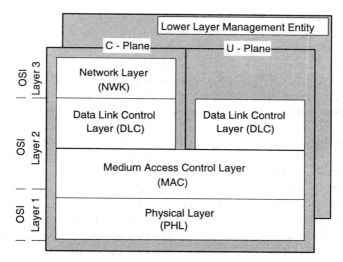

**Figure 3.8** DECT structure in relation to the OSI model

These two functions are used to create three services; a broadcast service, a connection orientated service (e.g. for telephony) and a connectionless service (e.g. for packet-like transmission).

The broadcast service is always transmitted from every base in a reserved data field (A field) on at least one physical channel. This beacon transmission allows portable parts (PP) to quickly identify and lock on to any suitable base (fixed part, FP) without requiring PP transmissions. The Data Link Control Layer (DLC) is largely responsible for providing very reliable data links to the network layer. This layer has two operational planes. The C-plane is concerned mainly with the fully error controlled transmission of internal control and signalling information. The U-plane offers a similar function in support of the specific requirements of the services being conveyed. For example the transparent unprotected service used for speech transmission.

Finally the Network Layer (NWK) is the main signalling layer of the protocol. It functions by exchanging messages between peer entities in support of, for example, establishment, maintenance and release of calls. Many additional messages support a range of independent capabilities and services. One group contains the necessary procedures that support cordless mobility which includes FP and PP authentication and location registration.

The lower layer management entity is concerned with procedures that involve more than one layer and yet are often only of local significance. Consequently they are only defined in general terms.

### 3.4.2 Interworking Units (IWU)

The transmission of information to end users beyond the DECT link requires additional protocols that are outside the DECT specification. Thus to interface a DECT link with, say, a GSM fixed part will require an appropriate IWU to establish proper unambiguous message transfer and in the process influence the service standard to be offered. Clearly the IWU concept will play a very important role in the full exploitation of the DECT specification.

### 3.4.3 Spectrum resource

Throughout Europe the band 1880MHz to 1900MHz has been set aside for use by the DECT system. This 20MHz of spectrum must however be used efficiently and flexibly if it is to meet the requirements of high capacity business systems. It is the main purpose of the Physical Layer to bring this about by ensuring that adequate capacity radio channels are created in a manner that permits high orders of radio channel re-use. The DECT spectrum resource has been distributed in space, frequency and time.

Spatial distribution is brought about because DECT supports the use of the well-known concept of cellular radio channel re-use. In this process the area to be served is covered by a number of base stations, each of which provides radio coverage over a limited radius. This radius is typically of the order 20–50 metres depending partly on the construction nature of the building being served but more likely on the density of telecommunications traffic to be catered for. In this latter case each small cell is able to offer a certain number of radio channels and the smaller the cell the shorter the distance away the same channels can be re-used with acceptable co-channel interference ratio. This is the classical cellular frequency re-use concept that gives very high orders of spectrum efficiency, expressed in terms of Erlangs per MHz per sq. km.

Frequency distribution is achieved by segmenting the available band into ten carrier frequencies from 1881.792MHz to 1897.344MHz and separated by 1728kHz.

Time distribution has been achieved by employing time division multiple access (TDMA) coupled with time division duplex transmission (TDD) to provide two-way communication on the same carrier frequency.

### 3.4.4 Detail radio aspects

The above applications place a number of requirements on the basic design of the radio system. The following are indicative of this process:

1. Ability to handover between channels on the same base station and between channels on different base stations without disturbance to speech or data.
2. Ability to offer variable bit rate channels to data and ISDN options.
3. Ability to cater for the traffic requirements of high capacity business communications systems.

The result of putting all these requirements into practice, including those of Section 3.2, produced an FDMA/TDMA/TDD system having ten carriers each conveying 12 time division duplex 32kbit/s channels (see Figures 3.9 and 3.10). To achieve the variable bit rate capability DECT is able to concatenate individual 32kbit/s timeslots to build up larger capacities. These concatenated timeslots need not be adjacent or on the same carrier frequency. Potentially the maximum throughput is, therefore, $12 \times 32 = 384$kbit/s both way. Handover is another key specified feature of DECT and has been designed to support handovers in 10ms to 15ms.

Other radio parameters are given in Table 3.2.

102 DECT digital specification

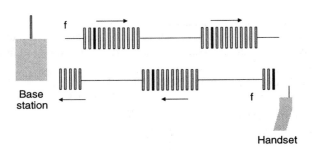

**Figure 3.9** DECT system. Each one of 10 carriers can serve up to 12 handsets or terminals

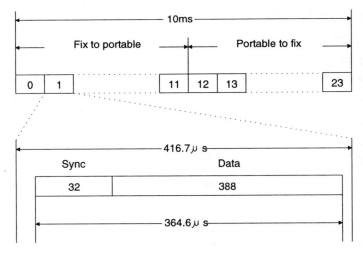

**Figure 3.10** DECT frame format

Each frame consists of 5ms alternate periods of transmission and reception and it is a general requirement that all base and portable terminals shall be able to operate on all timeslots.

This technical specification indicates that in cluttered, multi-path, environments outside buildings time dispersion may be a problem.

**Table 3.2** DECT radio parameters

| Parameter | Value |
| --- | --- |
| Frequency band | 1880–1900MHz |
| Radio channel spacing | 1.728MHz |
| Transmitted symbol rate | 1.152Mbit/s |
| R.F. carriers | 10 |
| Peak transmit power | 250mW |
| Modulation | Gaussian FSK (B.T. = c 0.5) |
| Nominal peak deviation | 288kHz |
| Duplex channels per frame | 12 |
| Frame period | 10ms |
| System loop delay | c 25ms |
| Bits per timeslot | 480 |
| Bits per burst | 420 |
| Guard time bits | 56 |
| Net channel rates (traffic) | 32kbit/s per slot (B field) |

Antenna diversity located at the base station is considered the first line of defence in such instances.

### 3.4.5 Radio operational features

The fixed part (FP) has an active channel for paging and synchronisation. An incoming call is paged on all FP and the PP responds on a channel chosen by the PP.

If channel quality deteriorates, PP moves out of range or due to increased interference, then handover is initiated to a more appropriate FP. A base controller synchronises this re-routeing of calls. PP periodically scan all channels to update their list of free channels to ease handover.

Link quality is assessed from channel bit error ratios and received signal strength.

For handover the PP requests a second channel (perhaps on another FP) to be modulated with the wanted signal. The PP then decides when to switch channels and then informs the controller which channel to release.

### 3.4.6 Frame structure

Figure 3.11 (ETSI, 1992b) shows the multiplex burst as it would appear for a normal telephony call and shows also the relationship between the PHL and MAC layer responsibilities. The 48 bits of the A field support the associated signalling channel requirements (C), the broadcast beacon transmission (Q) and the paging channel (P). The remaining 16 bits of the A field is a cyclic redundancy code (CRC) check used to protect the data.

The B field is used to transport the traffic information (I) and offers 320 bits per burst which is equivalent to 32kbit/s. The four X bits (which optionally can be extended to eight) are provided to detect collisions between bursts emanating from independent, and hence un-synchronised, systems. It should be noted of course that other multiplex structures are used in DECT particularly in the channel set up phase.

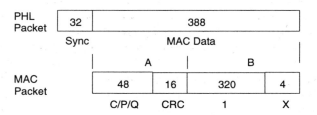

**Figure 3.11** PHL and MAC Layer responsibilities

## 3.4.7 Transmission plan and digital codec

DECT is subject to the same constraints as applied to CT2/CAI in terms of transmission performance. It is notable that both use the same codec standard and hence offer the same basic speech quality level.

However, since DECT employs TDMA working the inevitable consequence is that DECT loop delay amounts to about 25ms. Such a delay cannot be tolerated in the local public network in the presence of speech echoes from the DECT handset.

At present two approaches are being taken towards the amelioration of this effect:

1. By designing handset with virtually zero echo return. This calls for high orders of acoustic coupling loss between handset ear and mouth pieces approaching 46dB.
2. A combination of good handset design combined with the use of acoustic echo cancellation or control.

The latter seems to be preferred. Unfortunately the problem of echo cancellation and control in terminal equipment is not an independent design exercise as care needs to be taken to ensure that there are no unfavourable reactions with similar devices already in place on long distance connections, e.g. over a satellite link.

## 3.4.8 Implementation

To achieve success the basic production cost of DECT must be low. Thus the same objectives that were seen to be important to CT2/CAI's cost basis are equally applicable to DECT, i.e. the adoption of common components and large scale production to achieve the full cost benefits of scale.

Furthermore, equipment installation practices must be developed that can guarantee standards of communication quality that the customer has come to expect from the more conventional wired terminal equipment.

# 3.5 References

ETSI (1991) *Common Air Interface Specification to be used for Interworking between Cordless Telephone Apparatus in the Frequency Band 864.1MHz to 868.1MHz, Including Public Access Services. (prI-ETS 300 131).*

ETSI (1992a) *DECT Services and Facilities Requirements Specification*, ETSI Technical Report.

ETSI (1992b) *Digital European Cordless Telecommunications Common Interface. (prETS 300 175-1 to 9).*

ETSI (1992c) *Digital European Cordless Telecommunications Reference Document*, ETSI Technical Report.

Evans, M. W. (1990) CT2 Common Air Interface, *British Telecom Technical Journal*, **9**, (2), pp. 103–111.

Greenwood, D. (1995) Europeans decked by Japan's PHS, *Mobile Europe*, May.

Hendy, J. (1994) The case for DECT, *Telecommunications*, March.

Holmes, D. W. J. and Swain, R. S. (1990) The Digital Cordless Telecommunications Common Air Interface, *British Telecom Technical Journal*, **8**, (1), pp. 12–18.

Koopmans, H. (1995) Digital cordless: the worldwide rollout, *Telecommunications*, March.

MTN (1995) All hands on DECT, *MTN*, April/May.

Ochsner, H. (1990) Radio Aspects of DECT. In *Proceedings of the Fourth Nordic Seminar on Digital Mobile Radio Communications (DMR IV)*, 26–28 June, Oslo, Norway.

Rashidzadeh, B. and O'Connel, T. (1995) GSM and DECT — a dual mode solution, *Mobile Communications International*, April.

Swain, R. S. (1984) Cordless Telecommunications in the UK. In *Proceedings of the National Communications Forum XXXIII*, Chicago, USA, Sept.

Swain, R. S. (1985) Cordless Telecommunications in the UK, *British Telecom Technical Journal*, **3**, (2).

Swain, R. S. (1990) Digital Cordless Telecommunications — CT2, *British Telecommunications Engineering*, **9**, (2), pp. 98–102.

# 4. Cellular radio systems

## 4.1 Introduction

Cellular radio systems are by far the most common of all public mobile telephone networks, the earlier (pre-cellular) networks now all being in decline. The basic principles of cellular systems were established by Bell Laboratories in 1949, but it was not until the early 1980s that technology allowed real commercial networks to be built and service offered to the public.

Systems were developed at different times in different countries and subject to a variety of different constraints such as frequency band, channel spacing etc. As a result, a number of different and incompatible cellular standards are in use throughout the world, and the more important standards are summarised later in this chapter.

Work is already well in hand to specify and develop second generation cellular systems, for which the opportunity is being taken to develop common standards and systems across several countries. One such notable system, GSM, has been developed in Europe, and is described in some detail later in this chapter.

## 4.2 Principles of operation

### 4.2.1 Network configuration

In a cellular radio system, the area to be covered is divided up into a number of small areas called cells, with one radio base station (BS) positioned to give radio coverage of each cell. Each base station is connected by a fixed link to a mobile services switching centre (MSC), which is generally a digital telephone exchange with special software to handle the mobility aspects of its users. Most cellular networks consist of a number of MSCs each with their own BSs, and

## 108 Principles of operation

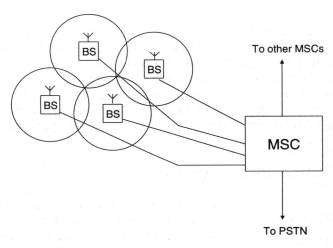

**Figure 4.1** Cellular network configuration

interconnected by means of fixed links. The MSCs interconnect to the public switched telephone network (PSTN) for both outgoing calls to, and incoming calls from fixed telephones. Figure 4.1 shows a typical network arrangement.

A cellular network will be allocated a number of radio frequencies, or channels, for use across its coverage area, this number being dependent upon the amount of spectrum made available by the licensing authority and the channel spacing of the technical standard used by the network. The radio channels are grouped together into a number of channel sets, and these sets are then allocated to the cells, one set per cell, on a regular basis across the whole coverage area. Each channel will therefore be re-used many times by the network. The method of radio planning and allocation of channels to the cells is described later in this chapter.

### 4.2.2 Signalling

Generally, one radio channel is set aside in each cell to carry signalling information between the network and mobile stations. In the land to mobile (L–M) direction, overhead information about the operating parameters of the network, including an area identifier code, is broadcast to all mobiles located in the cell's coverage area. In addition specific commands are transmitted to individual mobiles in order to control call setup and mobiles' location updating.

In the mobile to land (M–L) direction, the signalling channel is used by the mobiles to carry location updating information, mobile originated call setup requests, and responses to land originated call setup requests.

### 4.2.3 Location registration

When a mobile is not engaged in a call, it tunes to the signalling channel of the cell in which it is located and monitors the L–M signalling information. As the mobile moves around the network, from time to time it will need to retune to the signalling channel of another cell when the signal from the current cell falls below an acceptable threshold.

When the mobile retunes in this way, it reads the overhead information broadcast by the new cell and updates the operating parameters as necessary. It also checks the location information being broadcast by the new cell and, if this differs from the previous cell, the mobile automatically informs the network of its new location by means of an interchange on the signalling channel (Figure 4.2). By means of this location registration procedure, the network is able to keep updated a database of the location area of all mobiles. This information is used in the call set up procedure for land to mobile calls.

### 4.2.4 Call set up

The signalling procedures for mobile to land (M–L) and land to mobile (L–M) call set up depend upon the technical standard of the particular network. However the general procedure described below holds true for many networks. When the user wishes to make a call, the telephone number to be called is entered followed by a 'call initiation' key (e.g. pressing the SEND button).

110  Principles of operation

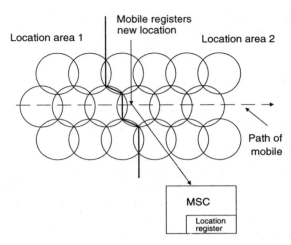

**Figure 4.2**  Mobile location registration

The mobile will transmit an access request to the network on the M–L signalling channel; this may be preceded by the mobile rescanning to ensure it is operating on the signalling channel of the nearest base station. If the network can process the call the base station will send a voice channel allocation message which commands the mobile to switch to a designated voice channel, namely one of the channels allocated to that cell. The mobile retunes to the channel indicated and the network proceeds to set up the call to the desired number. As part of the call set up procedure, the network will validate the mobile requesting the call to ensure that it is a legitimate customer. Many networks incorporate specific security features to carry out this validation.

When the network receives a call for a mobile (eg from the PSTN) it will first check the location database to determine in which location area the mobile last registered. Paging calls to the mobile are transmitted on the L–M signalling channels of all the base stations in the identified location area and a response from the mobile awaited. If the mobile is turned on and receives the paging call it will acknowledge to its nearest base station on the M–L signalling channel. The base

## 4.2.5 In-call handover

At all times during a call (whether L–M or M–L) the base station currently serving the mobile monitors the signal (strength and/or quality) from the mobile. If the signal falls below a predesignated threshold, the network will command neighbouring base stations to measure the signal from the mobile (Figure 4.3(a)). If another base station is receiving the mobile with a stronger signal than the current base station, a signalling message is sent to the mobile on the voice channel from the current base station commanding the mobile to a new voice channel, namely a free voice channel from those allocated to the neighbouring cell. The mobile changes frequency (and thereby the serving base station) and simultaneously the network connects the call to the new base station (Figure 4.3(b)).

The measuring process and new cell selection may take several seconds, but the user will only be aware of a brief break in transmission as the mobile tunes to the new voice channel.

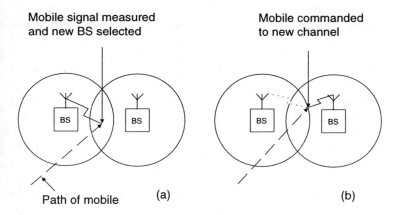

**Figure 4.3** In-call handover

### 4.2.6 Power control

Since the size of a cell may be anything from one kilometre to tens of kilometres across, it is not necessary for a mobile to transmit on full power at all times in order to maintain a satisfactory signal level at the base station's receiver. Most cellular standards therefore incorporate mobile power control, the base station commanding the mobile to transmit at one of a number of power levels. As the mobile moves closer to or further from the base station, further commands are issued to keep the received signal level to prescribed limits. By reducing the average mobile power level, co-channel interference is reduced, improving overall system quality.

## 4.3 Radio planning

As previously described, cellular radio re-uses the same radio channels in different cells and because of this re-use, two mobiles using the same channel in different cells may interfere with each other, a phenomenon known as co-channel interference. The key objective of planning a cellular radio system is to design the cell repeat pattern and frequency allocation in order to maximise the capacity of the system whilst controlling co-channel interference to within acceptable limits.

### 4.3.1 Cell repeat patterns

The cell plan has to be chosen such that the number of channel sets (N) fit together in a regular fashion without gaps or overlaps. Only certain values of N achieve this, and typical arrangements of interest to cellular radio are $N = 4$, 7 and 12 as shown in Figure 4.4. The value of N has a major effect on the capacity of the cellular system.

As the number of channels sets is decreased, the number of channels per cell increases, hence the system capacity increases. For example, if there are a total of 140 channels available, a 4 cell repeat pattern would provide 35 channels per cell, whilst a 7 cell would provide 20 channels per cell.

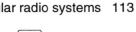

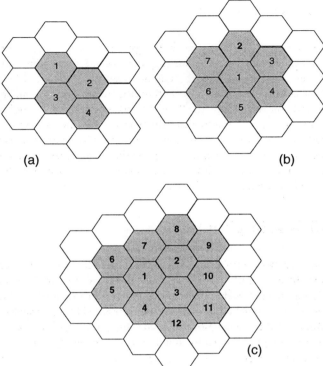

**Figure 4.4** Cell repeat patterns: (a) four cell repeat; (b) seven cell repeat; (c) twelve cell repeat

On this basis, the smallest possible value of N seems desirable. However, as N decreases, so the distance between cells using the same channels reduces, which in turn increases the level of co-channel interference.

The repeat distance D and the cell radius R are both related by the geometry of the cell pattern. These are shown in Figure 4.5 and Equation 4.1.

$$Re\text{-}use\ ratio\ \ D/R\ =\ \sqrt{3N} \qquad (4.1)$$

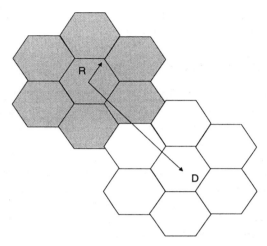

**Figure 4.5** Frequency re-use : D/R ratio

In practice, in a real network, it is not possible to achieve a regular cell pattern. This is because radio propagation at the frequencies used by cellular radio systems is affected by the terrain and by buildings, trees and other features of the landscape.

### 4.3.2 Co-channel interference

Generally, a mobile will receive a wanted carrier signal (C) from the base station serving the cell in which it is located, and in addition, interfering signals (I) from other cells. The carrier to interference ratio C/I is related to the re-use ratio D/R. Cellular radio systems are designed to tolerate a certain amount of interference, but beyond this, speech quality will be severely degraded. The TACS cellular system, for example, will work with a C/I down to around 17dB. This lower limit on C/I effectively sets the minimum D/R ratio that can be used.

The two key factors in ensuring that good quality transmission can occur between a mobile and base station are that the wanted signal strength is sufficiently large, that is, above the receiver threshold sensitivity, and that the interference level is low enough to give an adequate C/I ratio. Both of these factors depend on the radio propagation between the mobile and base stations.

### 4.3.3 Radio propagation

There are a number of elements which contribute to the received signal strength at a mobile. Firstly, for a line of sight path, there is a free space path loss which is related to the radial distance between base station and mobile.

In addition to this loss, where there is no direct line of sight path, there will be a diffraction loss resulting from obstructions in the path. In general there will also be an effect due to multiple signals arriving at the mobile due to reflections from buildings and other terrain features. This multi-path effect will result in signals either adding constructively or destructively.

As a mobile moves around within a cell it will experience varying signal, as shown in Figure 4.6, due to these factors. Fast fading is caused by the multipath effect, and occurs with only a small move-

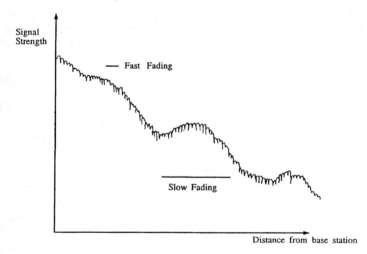

**Figure 4.6** Fading effects

ment of the mobile. This is also known as Rayleigh fading. Slow fading is mainly caused by terrain features and occurs over large distances of hundreds of metres.

In addition, the path loss is also dependent upon the type of terrain, for example, urban with dense buildings, or rural with trees, or even over water. The height of the mobile and base station above ground level also affects the propagation, although the mobile height is not usually a variable.

Predicting path loss is an essential part of radio planning, and because of the large number of contributing factors, empirically based formulae are used. The most widely used formula is the Hata model (Hata, 1980) which is based on the propagation measurement results of Okumura et al. (1968).

Hata's basic formula for the total path loss, Lp, is given by Equation 4.2 where $f_c$ is the carrier frequency in MHz, $h_b$ is the base station antenna height, $h_m$ is the mobile antenna height, R is the radial distance in kilometres, and $a(h_m)$ is the mobile antenna height correction factor.

$$Lp\,(dB) = 69.55 + 26.16 \log(f_c) - 13.82 \log(h_b) \\ - a(h_m) + (44.9 - 6.55 \log(h_b)) \log R \qquad (4.2)$$

Correction factors can be used to take into account the type of terrain.

### 4.3.4 Practical radio planning

Armed with a propagation model it is possible to calculate both the wanted signal strength and the interference level for all locations in a cell. Generally this is done using a computer based tool which can draw upon a database of cell site information and terrain data. Some advanced tools can also take account of diffraction losses. For practical purposes a planner will aim to achieve the required signal strength and C/I ratio over 90% of the cell coverage area, by varying antenna heights, transmitter powers, frequency allocations and other factors as appropriate.

**Table 4.1** Typical power budget (TACS)
(1 = Key planning parameters)

| Signal strength budget | Downlink | Uplink |
|---|---|---|
| Transmitter power (EIRP) | 50dBm | 39dBm |
| Receiver sensitivity (including antenna gain) | 106dBm | 113dBm |
| Fade margin (90% of area) | 4dB | 4dB |
| Required receiver level[1] | 102dBm | 109dBm |
| Maximum path loss (allowing for fading) | 152dBm | 148dBm |
| *Interference budget* | | |
| C/I threshold | 17dB | 17dB |
| Target C/I for planning (90% of area)[1] | 25dB | 25dB |

To simplify calculations, an allowance for Rayleigh fading and shadow fading is usually made within the system power budget. A typical power budget is shown in Table 4.1.

### 4.3.5 Adding capacity

Once a cellular network has been planned to provide overall coverage, there are a number of ways of adding additional capacity. A simple and cost effective option is to allocate further radio channels to existing cells. However, this can only be done by an extension band, for example the ETACS allocation in the UK. Other alternatives involve rearranging the cellular plan, either by cell splitting or by sectorisation.

Cell splitting is achieved by dividing an existing cell up into a number of smaller cells, by adding additional base stations as shown in Figure 4.7; it is then necessary to reallocate the radio channels. By repeatedly splitting cells; the cell size, and hence the system capacity,

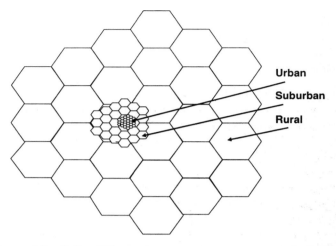

**Figure 4.7** Cell splitting

can be tailored to meet the traffic capacity requirements demanded by customer behaviour in all areas.

In rural areas, cells may be 20km to 30km in radius. In practice, as cell sizes decrease, propagation effects, particularly in city areas, cause an increase in co-channel interference, even if the repeat pattern is maintained. Also, as cell sizes decrease, it becomes increasingly difficult to find suitable base station sites, which need to be accurately positioned in order to keep to a regular pattern.

The cost of providing and maintaining a large number of individual base stations is also a factor, such that in addition to cell splitting, sectorisation of cells is commonly used in urban areas.

In a regular cellular layout, co-channel interference will be received from six surrounding cells which all use the same channel set. One way of cutting significantly the level of interference is to use several directional antennas at the base stations, with each antenna illuminating a sector of the cell, and with a separate channel set allocated to each sector.

There are two commonly used methods of sectorisation, using three 120 degree sectors or six 60 degree sectors as shown in Figure

Cellular radio systems 119

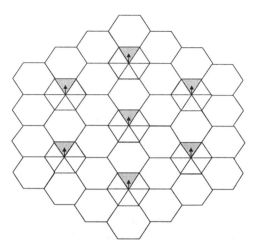

**Figure 4.8** Sectorisation

4.8, both of which reduce the number of prime interference sources to one. This is because, of the six surrounding co-channel cells, only one will be directed at the wanted cell.

A disadvantage of sectorisation is that the channel sets are divided between the sectors such that there are fewer channels per sector, and thus a reduction in trunking efficiency. This means that the total traffic which can be carried for a given level of blocking is reduced. However, this effect is offset by the ability to use smaller cells, such that the end result is a significant increase in total capacity.

## 4.4 Overview of systems

Cellular networks have been developed and deployed at various times and places in many countries across the world. In several cases, the local telecommunications authority or company was central to the specification and development of the standard with which the network complied. Since frequency allocations and other basic parameters (such as channel spacing) have often been set at national level

**Table 4.2** Comparison of system parameters

|  | AMPS | TACS | NMT 450 | NMT 900 | C-450 |
|---|---|---|---|---|---|
| Frequency band | 800 MHz | 900 MHz | 450 MHz | 900 MHz | 450 MHz |
| Channel spacing | 30kHz | 25kHz | 25kHz | 12.5kHz | 20kHz |
| Speech modulation | FM | FM | FM | FM | FM |
| Deviation | 12kHz | 9.5kHz | 5.0kHz | 5.0kHz | 4.0kHz |
| Signalling | Direct FSK | Direct FSK | Audio FFSK | Audio FFSK | Direct FSK |
| Signalling bit rate | 10 kbit/s | 8 kbit/s | 1200 bit/s | 1200 bit/s | 5280 bit/s |
| 'Overlay' signalling | NO | NO | NO | NO | YES |

and not co-ordinated between countries, these local factors resulted in different standards being adopted by different countries.

In addition, in cases where the development of a standard has started later, the opportunity has been taken to introduce new features made possible by advances in technology, further increasing the diversity between standards.

From this wide range of differing system standards, four have become largely dominant and have been adopted in many countries, albeit still with some variations. These four are AMPS, TACS, NMT (both NMT 450 and NMT 900) and C450, and their basic system parameters are shown in Table 4.2. Each of these is described in outline below, but in addition special mention must be made of the European designed GSM system which is set to become the dominant standard for Europe in the mid to late 1990s.

## 4.4.1 AMPS

AMPS stands for Advanced Mobile Phone System and was developed in the USA primarily by Bell Laboratories as a successor for the heavily congested IMTS (Improved Mobile Telephone System). Being designed for the North American market, AMPS uses the 800MHz band allocated to mobile services in ITU Region 2 (the Americas), with 30kHz channel spacing in common with established PMR practice.

AMPS uses analogue FM for speech transmission, but with a wider frequency deviation (12kHz) than is the norm for a 30kHz channelling system. By adopting the wide deviation, the dynamic range of the speech channel is extended and protection against co-channel interference is increased. This, together with the use of speech compression/expansion (companders) yields a high quality voice circuit with the capability to maintain performance in a high capacity (poor interference ratio) configuration.

Signalling between mobile and base station is at 10kbit/s, with Manchester encoding applied taking the bit rate to 20kbit/s. The data is modulated onto the radio carrier by direct frequency shift keying (FSK). Error control is achieved by multiple repetition (5 or 11 times) of each signalling word, with majority voting applied at the receiver to correct errors. A BCH block code is also applied to detect any uncorrected errors.

Whilst a call is in progress, the base station transmits a low level supervisory audio tone (SAT) in the region of 6kHz. Three different SAT frequencies are used by the network, and are allocated to the base stations so that the nearest co-channel base stations (i.e. those most likely to cause interference) have a different SAT from the wanted base station. The mobile continuously monitors the received SAT and also transponds the signal back to the base station. If the mobile (or the base station) detects a difference between the received SAT and that expected, the audio path is muted to prevent the interfering signal from being overhead. If the condition persists, the call is aborted.

AMPS underwent a long development period, and an extended trial (technical and commercial) which not only fixed the system parameters but also contributed to the basic planning rules which hold true for all cellular systems. The system design was comprehensively described in 1979 (Bell, 1979), but it was not until 1983 that operating licences were issued and true commercial exploitation of the system commenced.

AMPS is in operation extensively across the North American continent (USA and Canada). Due to the regulatory conditions in force in the USA, deployment has been in the form of a patchwork of largely independent standalone systems, with two competing systems operating in each licence area.

Although commercial roaming agreements exist to allow customers of one operating company to obtain service from another when they are in a different part of the country, a seamless nationwide service is, as yet, not available to the customer.

AMPS is now also used in a number of a number of Central and South American countries, in Australia and some Far East countries. World-wide it is the dominant standard in terms of installed customer base.

AMPS is being further developed to incorporate digital speech encoding, with TDMA techniques to give three digital voice channels per one radio channel. Digital AMPS (DAMPS) has the same basic architecture and signalling protocol as AMPS and is therefore more evolutionary than revolutionary (as is GSM in Europe).

### 4.4.2 TACS

TACS stands for Total Access Communications System, and was adapted from the AMPS standard by the UK when cellular radio was licensed for operation from 1985. The adaptation was necessary to suit European frequency allocations which were at 900MHz, with 25kHz channel spacing. This meant a reduction in frequency deviation and signalling speed was necessary (BS, 1990).

The signalling scheme of AMPS was retained largely unchanged, but some enhancements were introduced, particularly in the procedures for location registration, to make the standard more suitable for deployment in systems offering contiguous nationwide coverage. The opportunity was also taken to introduce extra features, such as signalling of charge rate information (e.g. for payphones).

TACS was originally specified to use the full 1000 channels (2 × 25MHz) allocated to mobile services in Europe. However in the UK, only 600 channels (2 × 15MHz) were released by the licensing authority, the remainder being reserved for GSM. Subsequently an additional allocation of channels below the existing TACS channels was made, namely the Extended TACS (ETACS) channels, and the standard was modified accordingly.

TACS equipment availability and cost have both benefitted from the standard's similarity to AMPS, and TACS systems have been adopted by several European countries (UK, Eire, Spain, Italy, Austria and Malta), in the Middle East (Kuwait, UAE and Bahrain) and the Far East (Hong Kong, Singapore, Malaysia and China). In Europe, TACS is on an equal footing with NMT in terms of installed customer base. A variant of TACS (called J-TACS) has also been adopted in Japan.

### 4.4.3 NMT

NMT stands for Nordic Mobile Telephone (system), and was developed jointly by the PTTs of Sweden, Norway, Denmark and Finland during the late 1970s/early 1980s. The system was designed to operate in the 450MHz band, and was later adapted to also use the 900MHz band. Although NMT was developed after AMPS, it saw commercial service before it, opening in late 1981.

NMT450 uses a channel spacing of 25kHz, speech modulation being analogue FM with a peak frequency deviation of 5kHz, the same as standard PMR practice.

NMT900 also uses a frequency deviation of 5kHz, but with a 12.5kHz channel spacing to double the number of available channels, albeit with a degraded adjacent channel rejection performance which must be taken into account during frequency planning. Signalling is at 1200 bit/s using audio fast frequency shift keying (FFSK). Error protection of the signalling information is by means of a Hagelbarger convolutional forward error correcting code.

NMT was designed from the outset to support international roaming and was first implemented with full four nation roaming in the four participating countries (Norway, Sweden, Finland and Denmark). Since then NMT450 has been deployed in many other European countries (Austria, Spain, Netherlands, Belgium, Luxembourg, France, Iceland, Faroe Is., Turkey and Hungary) but due to differences in the frequency allocations in the 450MHz band between countries, not all networks are fully compatible to allow roaming.

NMT900 was developed as a necessity as capacity became exhausted on the NMT450 networks, and has been deployed since 1987 as an overlay network in several countries, and in Switzerland as their main network.

### 4.4.4 C450

C450 (also known as Netz-C) was developed by Siemens during the early 1980s under the direction of the (West) German PTT, Deutsche Bundespost. Commercial service opened in 1985 following a trial period. C450 has a channel spacing of 20kHz, in common with other mobile services in Germany at 450MHz and speech modulation is analogue FM with a frequency deviation of 4.0kHz. Signalling for call control is transmitted at 5.28kbit/s by direct FSK. Error protection of the signalling is by bit interleaving with a BCH block code backed by an acknowledgement protocol.

In addition, C450 uses continuous signalling between base station and mobile during a call, achieved by time compressing the speech in bursts of 12.5ms, each burst being compressed into 11.4ms. This process opens up slots of 1.1ms duration every 12.5ms and the signalling data is inserted into these slots and extracted by the receiver which also time expands the speech back to its original form.

This continuous signalling serves several purposes:

1. It allows the base station to send power control and handover messages to the mobile without disturbing the voice channel.
2. The data is checked for jitter, and thereby the quality of the channel can be determined in order to indicate the need for a handover.

3. The time delay between a base station transmitting a data burst and receiving the response from the mobile is measured at the base station and used to calculate the distance between them. This distance is also taken into account in handover determination.
4. The data is used as a timing reference by the mobile to lock its internal clocks.

C450 contains a number of advanced features made possible by the application of current developments in technology. Although speech transmission is analogue, it can be regarded as a hybrid technology system, and several of its characteristics such as time slotted signalling channels and continuous signalling during call have been carried through into the GSM system design.

Coming later to the European scene, C450 has chiefly only served the German market, although systems are also operating in Portugal and South Africa.

## 4.4.5 GSM

The GSM standard was developed as a joint initiative by the members of the Conference of European Posts and Telecommunications administrations (CEPT) with the eventual aim of building a unified pan-European network, giving the user a near uniform service throughout all European countries. An added bonus of a common standard should be lower terminal equipment prices through economies of scale (Dutson, 1995; Josifovskas, 1995; Shetty, 1995).

Work on the standard started in 1982, and by 1987 all the basic architectural features were decided. The full Phase 1 specification was completed in 1990, but work continues on further phases incorporating new features and services.

In 1987, the majority of operators participating in GSM signed a Memorandum of Understanding (MoU) committing them to make GSM a reality by installing networks and opening commercial service by 1991. Since that time further operators have signed the MoU, bringing the total to date to 25.

The GSM technical standard makes full use of currently available levels of technology, incorporating features such as low bit rate speech, convolutional channel coding with bit interleaving and frequency hopping. The standard is intended to endure for many years to come.

## 4.5 Detailed description of GSM

### 4.5.1 GSM architecture

The basic architecture of GSM is not dissimilar to other cellular radio systems and comprises base transceiver stations (BTS), Base Station Controllers (BSC), Mobile Switching Centres (MSC), a variety of registers and a network management system, as shown in Figure 4.9.

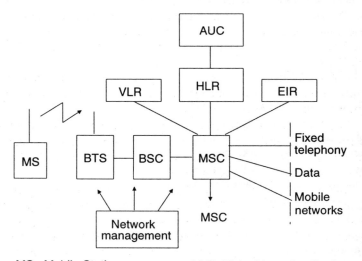

MS  Mobile Station
MSC  Mobile Switching Centre
BTS  Base Transceiver Station
BSC  Base Station Controller

VLR  Visited Location Register
HLR  Home Location Register
EIR  Equipment Identity Regsiter
AUC  Authentication centre

**Figure 4.9** GSM architecture

The mobile station comprises a mobile equipment and a subscriber identity module (SIM).

In addition to these functional entities, GSM also defines several interfaces, the Radio Interface (Um), the interface between the MSC and BSC (A interface) and the signalling interface which allows roaming between networks. This is based on the ITU-T No.7 signalling standard and is defined as a Mobile Application Part (MAP).

The BTS and BSC together form the Base Station Subsystem (BSS) and carry out all the functions related to the radio channel management.

This includes the management of the radio channel configurations, allocating radio channels for speech, data and signalling purposes, and controlling frequency hopping and power control. The BSS also includes, as does the MS, the speech encoding and decoding, and channel coding and decoding.

The MSC, VLR and HLR are concerned with mobility management functions. These include authentication and registration of the mobile customer, location updating, and call set up and release. The HLR is the master subscriber database and carries information about individual subscribers numbers, subscription levels, call restrictions, supplementary services and the current location (or most recent location) of subscribers.

The VLR acts as a temporary subscriber database for all subscribers within its coverage area, and contains similar information to that in the HLR. The provision of a VLR means that the MSC does not need to access the HLR for every transaction.

The authentication centre (AUC) works closely with the HLR and provides information to authenticate all calls in order to guard against fraud. The equipment identity register (EIR) is used for equipment security and validation of different types of mobile equipment. This information can be used to screen mobile types from accessing the system, for example if a mobile equipment is stolen, not type approved, or has a fault which could disturb the network.

Network management is used to monitor and control the major elements of the GSM network. In particular, it monitors and reports faults and performance data. It can also be used to re-configure the network.

## 4.5.2 Air interface

The GSM Air Interface (Um) provides the physical link between the mobile and the network. Some of the key characteristics of the air interface are given in Table 4.3. As already described, GSM is a digital system employing time division multiple access (TDMA) techniques and operating in the 900MHz band. The CEPT have made available two frequency bands to be used throughout Europe by the GSM system, namely:

1. 890MHz to 915MHz for the mobile to base station (uplink) direction.
2. 935MHz to 960MHz for the base station to mobile (downlink) direction.

These 25MHz bands are divided into 124 pairs of carriers spaced by 200kHz. In addition, consideration is now being given to specifying additional carriers in a pair of extension bands 872MHz to 888MHz and 917MHz to 933MHz. Each of the carriers is divided up

**Table 4.3** GSM air interface parameters

| Frequency band mobile – base | 890 – 915MHz |
|---|---|
| Frequency band base – mobile | 935 – 960MHz |
| 124 radio carriers spaced by 200kHz ||
| TDMA structure with 8 timeslots per radio carrier ||
| Gaussian Minimum Shift Keying (GMSK) Modulation with BT = 0.3 ||
| Slow frequency hopping at 217 hops per second ||
| Block and convolutional channel coding with interleaving ||
| Downlink and uplink power control ||
| Discontinuous transmission and reception ||

into eight TDMA timeslots of length 0.577ms such that the frame length is 4.615ms. The recurrence of each timeslot makes up one physical channel, such that each carrier can support eight physical channels, in both the uplink and downlink directions.

The timeslot allocation in either direction is staggered so that the mobile station does not need to transmit and receive at the same time. Data is transmitted in bursts within the timeslots and a number of different types of burst can be carried as shown in Figure 4.10. The normal burst has a data structure as shown. It consists of 148 bits of which 114 are available for data transmission, 26 are used for a training sequence which allows the receiver to estimate the radio propagation characteristics and to set up a dispersion equaliser, 6 bits as tail bits, and two stealing flags. These physical channels therefore provide a data throughput of 114 bits every 4.615ms or 24.7kbit/s.

The bursts modulate one of the RF carriers using Gaussian Minimum Shift Keying (GMSK) modulation with a BT index of 0.3. The allocation of the carrier can be such that frequency hopping is achieved, i.e consecutive bursts of a physical channel will be carried by differing RF carriers. This "hopping" is performed every TDMA frame, or every 4.615ms and provides extra protection against channel fading and co-channel interference.

A number of logical channels can be carried by the physical channels described above. These are summarised in Table 4.4.

There are two categories of traffic channels; speech, whether full rate using 22.8kbit/s or half rate using 11.4kbit/s, and data, providing a variety of data rates. There are four basic categories of control channels, known as the broadcast control channel (BCCH), the common control channel (CCCH), the standalone dedicated control channel (SDCCH) and the associated control channel (ACCH). These are further divided into channels with specific purposes and for a detailed description of these channels the reader is referred to the GSM Recommendations published by ETSI.

Each of these logical channels is mapped onto the physical channels, using the appropriate burst type as shown in Figure 4.10.

TDMA frames are built up into 26 or 51 frame multiframes, such that individual timeslots can use either of the multiframe types, and then into superframes and hyperframes as shown in Figure 4.10. The

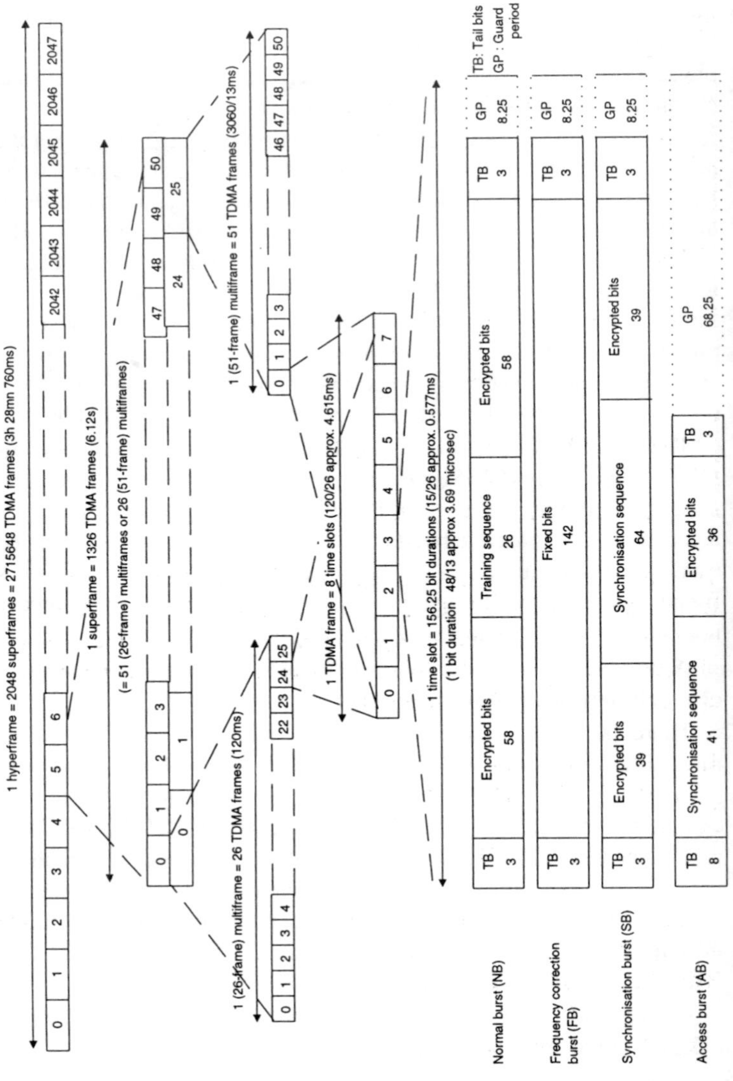

**Figure 4.10** GSM timeframes, timeslots and bursts. (Extract from GSM Recommendations 05.01)

# Cellular radio systems

**Table 4.4** GSM logical channels

| Traffic Channels (TCH) | | Control Channels (CCH) | | | |
|---|---|---|---|---|---|
| Speech | Data | Broadcast CCH (BCCH) | Common CCH (CCCH) | Standalone Dedicated CCH (SDCCH) | Associated CCH (ACCH) |
| Full rate TCH/F | TCH/F9.6 | Frequency Correction (FCCH) | Paging Channel (PCH) | | Fast (FACCH) |
| | TCH/F4.8 | Synchronisation (SCH) | Random Access (RACH) | | Slow (SACCH) |
| | TCH/F2.4 | | | | |
| Half rate TCH/H | TCH/H4.8 | | Access Grant (AGCH) | | |
| | TCH/H2.4 | | | | |

TCH and the associated ACCH uses the 26 frame structure, whilst the BCCH and CCCH use the 51 frame structure. The SDCCH may occupy one physical channel, providing 8 SDCCH, or may share a physical channel with the BCCH/CCCH. Typical arrangements for allocating the 8 physical channels could be:

1. 7 channels TCH and SACCH + 1 channel BCCH/CCCH/SDCCH
2. 6 channels TCH and SACCH + 1 channel BCCH/CCCH + 1 channel SDCCH.

Each cell must have at least one physical channel assigned to the BCCH/CCCH, where there are 2 or more carriers per cell, the non-BCCH carriers may have all 8 channels allocated to TCH.

## 4.5.3 Speech coding and channel coding

The speech coder is a regular pulse excited linear predictive coder (RPE-LPC) with long term prediction. This provides a net bit rate of

## 132 Detailed description of GSM

13kbit/s. It is a block based coder where the input samples are analysed in blocks with a 20ms duration. Work is also being carried out to specify a half rate speech coder which will effectively double the system capacity of GSM.

Before being assembled into the timeslots and frames, the digital speech and signalling data is encoded and interleaved. The speech coder output is divided up into three classes of bits and the most sensitive bits are encoded by adding parity check bits followed by a convolutional coder. Signalling data is encoded using a FIRE code. A process of interleaving is then used to spread the data blocks over a number of bursts.

For speech, an interleaving degree of 8 is used, i.e the speech block is spread over 8 bursts, whilst an interleaving degree of 4 is used for signalling. This overall process is shown in Figure 4.11, and the

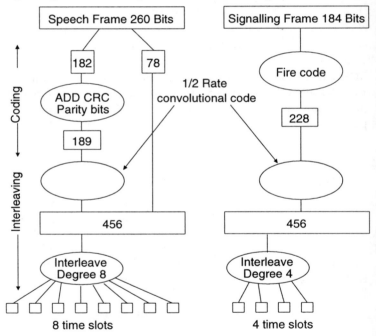

**Figure 4.11** GSM channel coding and interleaving

combined use of coding and interleaving provides good protection of channel data from the fading, dispersion and interference effects on the radio path. With the addition of frequency hopping and diversity techniques, the GSM air interface is particularly robust.

One of the penalties to be paid for this is the overall transmission delay. The speech coder contributes about 25ms and the channel coding and interleaving a further 37ms. The rest of the transmission delay budget allows for analogue to digital conversions, 16kbit/s transmission and switching in various parts of the network. The overall one way transmission delay thus amounts to around 90ms. Such a delay means that echo control is necessary even on short national calls.

### 4.5.4 GSM signalling

Figure 4.12 shows the overall signalling model. The Air Interface uses LAPDm Layer 2 signalling protocol and this is also used for the A-bis, BTS to BSC interface.

The layer 3 protocol consists of three sublayers, dealing with radio resource management (RR), mobility management (MM), and con-

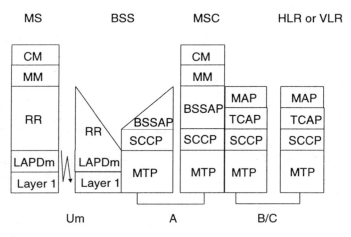

**Figure 4.12** GSM signalling model

nection management (CM). Radio resource management is concerned with managing the logical channels, including paging, channel assignments, handover, measurement reporting, and other functions.

The mobility management layer contains functions necessary to support the mobility of the user which include authentication, location updating, attach and detach of IMSI (International Mobile Subscriber Identity), and registration. The connection management layer is concerned with call control, establishing and clearing circuits, management of supplementary services and the short message service.

The BSC to MSC A-interface, and the various MSC to Register interfaces employ ITU-T No.7 signalling using the Message Transfer Part (MTP), Signalling Connection Control Part (SCCP), Transaction Capabilities Part (TCAP) and Mobile Application Part (MAP).

An example of the signalling messaging for establishing a mobile originated call is shown in Figure 4.13. The key events are:

1. Request and assignment of a channel, between MS and BSS.
2. A service request procedure which accesses the VLR.
3. An authentication and ciphering exchange which validates the mobile user and sets the encryption cipher.
4. Call set up which includes sending of dialled digits and establishing the connection.

Location updating is shown in Figure 4.14. An update request is indicated by the mobile and passed to the VLR in the new location area. The new VLR requests the IMSI from the old VLR and then signals the new location to the HLR. The HLR provides the subscriber data to the new VLR and cancels the subscriber entry in the old VLR. Finally a confirmation message is sent back to the mobile.

There are, of course, many other signalling exchanges, dealing with mobile terminating calls, supplementary services, and short message service. There is not space in this chapter to deal with the detailed signalling for these cases; the examples above describe the general principle and illustrate the roles of the MS, BSS, MSC, VLR and HLR.

# Cellular radio systems 135

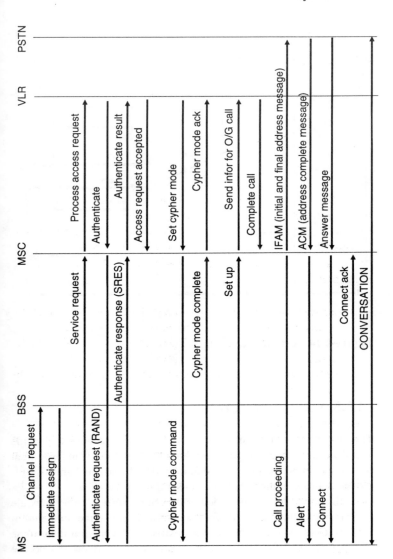

**Figure 4.13** Mobile originating call

## 136 Detailed description of GSM

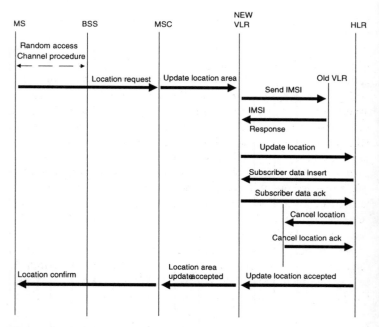

**Figure 4.14** Location updating

### 4.5.5 Security features

The information on the air interface needs to be protected, to provide user data (including speech) confidentiality and to prevent fraudulent use of subscriber and mobile identities. The basic mechanisms employed are user authentication and user data encryption.

Each mobile user is provided with a Subscriber Identity Module (SIM) which contains the IMSI, the individual subscriber authentication key (Ki) and the authentication algorithm (A3). After the mobile user has made an access and service request, the network checks the identity of the user by sending a random number (RAND) to the mobile. The mobile uses the RAND, Ki and A3 algorithm to produce a signed response SRES. This response is compared with a similar

response calculated by the network, and access only continues if the two responses match.

The SIM also contains a cipher key generating algorithm (A8). The MS uses the RAND and A8 to calculate a ciphering key (Kc) which is used to encrypt and decrypt signalling and user data information.

The authentication centre (AUC) is responsible for all security aspects and its function is closely linked with the HLR. The AUC generates the Ki's and associates them with IMSIs, and provides the HLR with sets of RAND, SRES and Kc for each IMSI. The HLR then provides the appropriate VLR with these sets and it is the VLR which carries out the authentication check. Authentication of mobile users can be carried out on call set up, both mobile originated and mobile terminated, on location updating, and on activation of supplementary services. As the authentication sets are used up in the VLR, further sets are requested from the HLR.

An additional security feature of GSM is the equipment identity register (EIR). This enables monitoring of the mobile equipment IMEI (International Mobile Equipment Identity) which is used to validate mobile equipments thus preventing non-approved, faulty or stolen equipment from using the system. This range of security features provide a high degree of protection to the user and the network operator.

### 4.5.6 GSM services and features

In addition to speech, GSM offers a wide range of data bearer services up to 9.6kbit/s suitable for connection to circuit switched or packet switched data networks. GSM also supports Group III facsimile as a data service by use of an appropriate convertor.

A comprehensive range of supplementary services are offered by GSM, including call forwarding, call barring, multi-party service, advice of charge and others. A full description is provided in the GSM Recommendations, and further detail of cellular services is provided later in this chapter.

An important feature of GSM is the short message service (SMS). This allows transmission of alphanumeric messages of up to 160 characters to or from a mobile via a service centre. If the message cannot be delivered due to the mobile being switched off, or outside of the coverage area, the message is stored at a service centre and re-transmitted when the mobile registers again. Received messages can be displayed on the mobile and stored in the SIM for future reference.

A related service is cell broadcast which allows messages of up to 93 characters to be sent to all mobiles within a specific geographical area, for example to deliver traffic or weather reports.

### 4.5.7 Roaming

Naturally, with a pan-European system, roaming of subscribers between networks is specified by GSM. When a mobile first switches on in a foreign PLMN (Public Land Mobile Network), the local MSC/VLR will determine the identity of the home PLMN from the mobile network code which is part of the IMSI.

The home HLR will be interrogated to establish whether roaming is permitted and for authentication. The home HLR then passes the subscriber data to the local (foreign) VLR and registers the foreign location of the mobile. Calls to and from the roamed mobile can then take place.

## 4.6 Services

The primary purpose of all cellular radio networks is to offer speech telephony service to its customers. In addition most networks offer a range of supplementary and value added services to enhance the basic product.

In analogue systems, basic telephony is provided directly by the audio path between mobile and network. Other than some linear speech processing to increase the channel's signal to noise performance, the audio path is transparent across the speech band, allowing other sounds (tones, non-voice signals etc.) to pass through undistorted. By contrast GSM (and other digital systems) use a speech coder tailored to voice characteristics. They therefore provide a fully acceptable telephony service, but non-voice signals can suffer distortion across the non-audio transparent path.

## 4.6.1 Supplementary services

Supplementary services are provided by means of enhancements to the basic call processing software in the MSCs. Many of these services have specific relevance to the cellular radio user, and in the main they parallel services which are becoming increasingly available on the fixed telephone networks (such as BT's Star Services in the UK).

Typical services are as follows:

1. Call divert, where all calls are diverted to the specified number, which may be another mobile or a termination on another network. This is of use if the user wishes to make calls but not receive them.
2. Divert on no answer, where calls are diverted to the specified number when the user does not answer within (for instance) 20 seconds. This is of use if a mobile is left switched on in an unattended vehicle.
3. Divert on mobile unavailable, where calls are diverted to the specified number if the network cannot contact the mobile owing to its being turned off or out of range. This is of particular use in a cellular system where, in general, users are not available at all times, and where coverage is not universal. This service is often combined with the 'divert on no answer' service.
4. Divert on busy, where calls are diverted to the specified number when the mobile is already engaged on a call. As an alternative, networks also provide call waiting.
5. Call waiting, where if a call is received when the mobile is already engaged on a call, the user is informed that a second call is waiting, and can choose to place the first call on hold whilst dealing with the second caller.
6. Three party calling, where the mobile user may set up calls to two other parties and connect them in a three way conference. This service can also be used to make enquiry calls whilst holding the original call.

## 4.6.2 Value added services

Value added services are normally provided by means of peripheral units attached to the cellular network, or to the fixed network with which to cellular network interconnects. In some countries, the prevailing regulatory regime will influence what services may be offered and in what manner, however, the following are typical.

### 4.6.2.1 *Messaging services*

Voice messaging is commonly available in association with cellular networks. Used in conjunction with the call divert supplementary services, the messaging service can pick up calls when the user cannot, and the caller can leave a message for later retrieval by the user. Some services allow the user to be alerted to the receipt of messages by means of a radiopaging service, or in some cases by a ringback on the cellular network itself.

In addition to voice messaging, GSM networks will incorporate the 'Short Message Service' which effectively turns a GSM mobile into a two way alphanumeric pager with forced message delivery and message delivery confirmation.

### 4.6.2.2 *Information services*

Voice information services are commonly available on fixed networks, normally carrying some premium call charge. Some services (such as travel and weather information) are of particular value to a mobile user and some networks make these more readily accessible, for instance by using the mobile's current location to select the appropriate information for that area.

### 4.6.2.3 *Private interconnect*

A large user of cellular can often gain economies by leasing a direct connection between the cellular network and their company's private network, thus bypassing the PSTN. Call charges for such direct connections are tariffed by the cellular operator at a level substantially less than that for PSTN calls. An extra benefit of private interconnect are that calls can be delivered direct to extensions on a company's network without having to be handled by the switchboard operator, saving time and labour.

### 4.6.3 Data services

In analogue cellular systems, the transparent audio path between the network and mobile can be used not only for voice communication, but also for non-voice communication such as data using in-band modems, and facsimile. In order to be used in conjunction with a mobile, data modems and fax machines which are designed for PSTN use have to be adapted for connection to the mobile by means of a special interface. Such interfaces are available for a range of mobiles, and often permit automatic call establishment and clear down under the control of the modem or fax machine.

The data rate achievable over a cellular radio channel will often be less than that over a direct PSTN path, mainly due to the more limited frequency response of the channel, and the delay spread characteristic which is affected by the audio processing in both mobile and base station. However data transmission at 1200bit/s (using ITU-T V.22) and 4800bit/s (using V32) can be achieved quite commonly on cellular networks, as well as fax up to 7200 or 9600bit/s.

The radio link between a cellular network's base stations and a mobile station is a notoriously hostile environment for data transmission. Disturbance and interruptions come from a variety of sources, such as variability of the radio signal strength, noise and interference, and 'intentional' breaks due to signalling interchanges between base station and mobile for handover and power control. In order to transmit data reliably over such a path, error control of some form is essential.

The simplest form of error control is a layer 2 protocol, and the emergence of the ITU-T V.42 standard has led to error correcting modems becoming readily available. Although V.42 (which contains two protocols, the 'open' LAP-M and 'proprietary' MNP4) was designed for fixed PSTN use, it has proved to perform sufficiently well over cellular paths, particularly to static mobiles, for the user to receive good service.

Many proprietary protocols have been specifically developed to cope with the errors experienced over cellular radio channels. One such protocol is called Cellular Data Link Control (CDLC), and was developed in the UK by Racal Vodata. CDLC uses two levels of error correction with dynamic switching, and techniques such as forward error correction, bit interleaving and BCH block coding with a basic HDLC protocol to give a highly robust data transmission path, even over poor quality channels.

Facsimile transmission over cellular has benefited by the increasingly widespread adoption of Group 3 error correcting (ECM) fax machines and the availability of portable machines suitable for vehicle use.

The GSM system does not provide a transparent audio path due to the voice coding techniques used, so data transmission in GSM is dealt with differently (Booth, 1995; Emmerson, 1995; Fenn, 1995; Franzon, 1995; Gross and Duffy, 1995; Murch and Stiffe, 1995; Postlethwaite, 1995). When the data mode is selected, the speech coder is replaced by a rate adaptor and channel coder which apply forward error correction to the data bits, and the resulting bit stream is then transmitted across the radio path in the same burst structure as for voice transmission. At the receive end the bit stream is extracted and errors are corrected up to the limit of the forward error correction scheme. If there are any errors remaining, a higher layer protocol is needed to detect and correct them.

GSM has defined two families of data services, termed transparent and non-transparent. The transparent service applies only forward error correction as described above, and the user application must be able to cope with the residual error rate. The characteristics of the transparent service are constant delay and throughput but variable error rate. The transparent service is of particular use in synchronous applications (eg X.25, IBM SDLC) where the higher layer protocol inherent in the application will correct the errors. Asynchronous applications may also use the transparent service, particularly at low bit rates where the forward error correction applied by GSM is stronger.

The non-transparent service applies a GSM specific layer 2 protocol between the mobile and the network in order to correct all residual errors, resulting in a near zero error rate. The penalty, however, is variable throughput and delay, dependent upon the prevailing radio conditions. The non-transparent service is of particular application to simple asynchronous terminals, although provision in the standards is also made for protocol conversion to allow X.25 packets to be carried.

Facsimile transmission over GSM is complicated by the use in the Group 3 standard of a number of data transmission rates and modem types (V.21, V.29, V.27). In order to carry the fax signals, GSM mobiles need a special adaptor to convert the multiple standards into a synchronous bit stream for transmission between mobile and network. A similar converter in the network then converts the signal back into the Group 3 protocol to interwork with fax machines in the fixed network.

## 4.7 Future developments

The technology of cellular radio systems continues to develop very rapidly. The early 1980s saw the introduction of the first commercial analogue systems and by the end of the decade trials of second generation digital systems were already under way. Systems such as GSM are now in service and work is already starting on the specification of a third generation world wide standard system.

These developments are not introducing technology for its own sake, but are aimed at improving the quality, capacity, and availability, and reducing the cost of mobile communications. In addition to these step changes in 'generations' of system there are technical advances which are applicable to current systems. These include techniques such as microcellular and intelligent networks.

### 4.7.1 Microcells

As the capacity of cellular systems has increased, cell sizes have decreased, in some networks to as small as 0.5km radius, such that controlling co-channel interface becomes a major problem. The use of microcells, that is, very small cells, is a way of increasing capacity still further. In a microcellular layout, base station antennas are placed below the building height in urban areas, and low power is used such that the propagation characteristics between base station and mobile are dominated by the street layout. Interference from adjacent cells is blocked by buildings.

Microcellular techniques allow significantly higher traffic densities to be achieved, and also enable smaller, lower power mobiles to be used.

The use of microcells requires improved handover techniques, which allow for fast and reliable handoff, for example when turning a street corner. One way of easing handover problems is to employ an 'umbrella cell' arrangement using conventional cells overlaying the microcells such that handover can be made into the umbrella cell where no suitable adjacent microcell can be identified. This also avoids the need to plan a contiguous coverage of microcells in an urban area.

New technology is now enabling the use of more compact and cheaper base stations. Conventional base sites have generally required a purpose built building, or rented space within an existing building for installation of base station racks of equipment. Now, base stations can be housed in small roadside or roof top mounted cabinets, and further reductions in size can be expected. Small base station equipment, and antennas, are essential to enable microcells to be built cost effectively.

## 4.7.2 Intelligent networks

Intelligent Network techniques (IN), are not, of course unique to cellular systems and have already become well established in fixed networks for the provision of 'free fone' or 'toll-free' type services, for example. However, the ability of an IN architecture to provide customised services is particularly valuable to a mobile user, who can have improved control over the handling of incoming calls. IN techniques also provide the ability to create a wide variety of advanced services.

Second generation cellular systems such as GSM are already designed around an architecture which can support IN type applications. In particular, the HLR function is closely related to the IN service control point. We can expect further developments in the near future which will bring a range of IN features to both the mobile user and the service provider.

### 4.7.3 Personal communications

The term PCN, Personal Communications Network, is used widely in the UK, whilst PCS, Personal Communications Services is used in the USA. Both aim at the same objective of serving the mass consumer market with mobile communications. The key challenge is to provide a very high capacity network to support a large number of users at low cost.

Microcellular techniques will certainly be needed, and in order to keep costs down, the concept of regional service, and local access to the PSTN is being considered. IN techniques may offer personal numbering across a variety of networks.

PCN is dealt with in detail in Chapter 5. The standard in Europe, known as DCS1800 is based on the GSM standard but operating at 1800MHz. There is therefore unlikely to be a significant technical difference between Cellular GSM and PCN, with microcellular techniques being equally applicable to either system.

In the USA, the use of CDMA, code division multiple access, is being trialled for PCS (Mobile, 1995). CDMA works on the principle of transmitting unique (orthogonal) codes to identify different users. Detection of signals is achieved by using correlating receivers such that other users appear as pseudonoise. CDMA thus allows a large number of users to share the same (wideband) radio channel.

There is considerable debate about the advantages and disadvantages of CDMA, in particular how to control near/far user interference; the extent to which this can be achieved is crucial to the ultimate capacity of CDMA. One of the key benefits of CDMA is the potential to share spectrum with other users, for example fixed links, and for this reason it is particularly attractive where additional spectrum for mobile systems cannot be made available.

## 4.8 Conclusion

Cellular radio is a comparatively young technology. Networks employing analogue systems have developed rapidly and now provide high quality service and excellent coverage in many of the developed countries. Technology developments are now increasing the potential network capacity, reducing the size of mobiles, and bringing advanced features and services to the mobile user. The decade ahead with the opportunity to introduce new digital systems and create a world-wide land mobile standard looks particularly exciting.

## 4.9 References

Bell (1979) Special issue on Advanced Mobile Phone Service, *Bell System Technical Journal*, January.

Booth, N. (1995) The quiet snowball, *Communications News*, January

BS (1990) Total Access Communication System (TACS) *BS6940 Parts 1 & 2*.

Dutson, B. (1995) A multi-billion $ opportunity, *Mobile Europe*, June.

Emmerson, B. (1995) Mobile data on a roll, *Mobile Communications International*, May.

Fenn, J. (1995) Channelling GSM data, *Mobile Europe*, September.

Franzon, G. (1995) Mobile data rolls on, *Telecommunications*, March.

Gross, D. and Duffy, R. (1995) World without wires, *Communications International*, June.

Hansen, S. (1988) Voice Activity Detection (VAD) and the Operation of Discontinuous Transmission (DTX) in the GSM System. In *Proceedings of the Digital Cellular Radio Conference*, 12–14 October, Hagen, Germany.

Hata, M. (1980) Empirical Formula for Propagation Loss in Land Mobile Radio Services, *IEEE Transactions*, **VT- 29**, (3), August.

Hodges, M.R.L. (1990) The GSM Radio Interface, *Br Telecom Technology Journal*, **8**, (1).

Josifovaks, S. (1995) The express through GSM country, *Electronics Weekly*, 31 May.

Mallinder, B.J.T. (1988) An Overview of the GSM System. In *Proceedings of the Digital Cellular Radio Conference*, 12–14 October, Hagen, Germany.

Mobile (1995) CDMA — the challenge ahead, *Mobile Europe*, September.

Murch, A. and Stiffe, P. (1995) Cellular data services over GSM, *Telecommunications*, March.

Okumura, Y. et al. (1968) Field Strength and its Variability in u.h.f. and v.h.f. Land Mobile Radio Service, *Review of the Electrical Communication Laboratory*, (6).

Oschner, H (1988) Overview of the Radio Subsystem. In *Proceedings of the Digital Cellular Radio Conference*, 12–14 October, Hagen, Germany.

Postlethwaite, D. (1995) Mobile data — a market on the move? *Mobile Communications International*, January.

Shetty, V. (1995) GSM: global superairway, maybe? *Communications International*, March.

Van der Arend, P.C.J. (1988) Security aspects and the implementation in the GSM System. In *Proceedings of the Digital Cellular Radio Conference*, 12–14 October, Hagen, Germany.

Vary, P. (1988) GSM Speech Codec. In *Proceedings of the Digital Cellular Radio Conference*, 12–14 October, Hagen, Germany.

# 5. Personal communication networks

## 5.1 Introduction

Personal Communications Networks or PCN is a unique concept in that a commercial requirement has driven the development of a telecommunications standard. This commercial requirement has resulted in the development of a standard within 18 months. Contrast this with ISDN where standards have been evolving for 10 years and only now are networks being introduced.

This chapter will explain what PCN is and why it came about.

## 5.2 History

The United Kingdom was one of the first countries to consider introducing PCN type services (Hadden, 1994) when its Department of Trade and Industry (DTI) outlined the concept of Personal Communications Networks in its paper 'Phones on the Move' in January 1989. This expressed the government's intention to allocate Radio Spectrum within the range 1.7GHz to 2.3GHz. The government held the idea that 'telephones should be for people and not places', whereby small personal communicators would allow access to the telecommunications network wherever that person should be.

The DTI required that any personal communications network should conform to a publicly available technical standard and as an early introduction was required using up-to-date technology, the view was expressed that pan-European Digital Cellular Technology would be a strong contender offering the following advantages:

1. Early implementation in line with pan-European digital systems.

2. Cost benefits arising from economies of scale.
3. The possibility of having customers' equipment which can operate on both PCN and pan-European Systems.
4. Staying in the mainstream of European developments and standards.

'Phones on the Move' concluded by seeking views on the following questions, which are relevant to wherever PCN type services are being introduced, as in North America:

1. Whether the market within the country will be able to support more public mobile radio operators?
2. Whether improvements could be made over today's networks with the advent of the pocket radio telephone?
3. The date by which new Personal Communication Networks could/should be implemented?
4. The most effective frequency. For example in Europe is the combination of 1.7GHz to 2.3GHz frequency channels and pan-European digital cellular radio technology the most effective solution for the early to mid 1990s?
5. How much bandwidth is needed for each operator in order to build up and maintain a viable subscriber base?
6. Should handover be a feature for a Personal Communications Network?
7. Should an immediate move to seize market opportunities be given priority over waiting to see where future international allocations for mobile services will be located?

Following the expression of views, the government allocated PCN licences to three consortia: Microtel, Unitel and Mercury PCN. All three companies decided to introduce PCN networks based on the ESTI GSM standard and expressed their intention to bring in service during 1992. Eventually the first commercial service was launched by Mercury One-2-One in September 1993, followed by Orange, which is owned by Hutchison Microtel, in April 1994 (Struthers Watson, 1994).

## 5.3 PCN definition

PCN is a system which provides the ability for customers to make and receive telephone calls from their pocket radio telephone: anytime, anywhere in the country. The system is based upon GSM, but operates within the 1.8GHz frequency band. It provides all the services you would expect from a modern digital telephone network. People can use the service in the car, or on the train, in fact anywhere they want to. It will offer ISDN-type services such as calling number identification, diversion, call back when free, call waiting, etc.

### 5.3.1 PCN and GSM

The personal communications networks in the UK are based upon GSM, but operate at 1.8GHz.

ESTI GSM was tasked with developing a version of GSM for operation in the 1.8GHz range. The series of recommendations became DCS1800 (Digital Cellular System at 1800MHz).

ESTI agreed to produce DCS1800 in two phases:

1. Phase 1 by January 1991 establishes the generic differences between GSM 900 and DCS1800.
2. Phase 2 establishes a common framework for both PCN and GSM.

For Phase 1 the changes to the GSM900 standards were:

1. Increase in bandwidth from 25MHz to 75MHz in each direction. This has impacts on the signalling and some RF aspects.
2. RF link definition to account for the translation from 900MHz to 1800MHz and to reflect the low power handsets.

So that the networks could be rolled out quickly, giving faster geographic coverage, the notion of National Roaming was introduced. This allows a number of network operators to share infrastructure.

Personal communication networks 151

To allow the RF parameters for DCS1800 to be calculated, six scenarios were examined:

1. Single MS (mobile station), single BTS (base transceiver station).
2. Multiple MS and BTS where operation of BTS's is co-ordinated (single operator).
3. Multiple MS and BTS where operation of BTS's is uncoordinated (multiple operator).
4. Collected MS.
5. Collocated BTS.
6. Collocation with other systems.

The impact of each of the above on radio performance was assessed. Practical worst case conditions were assumed. One critical parameter that had to be determined was the mobile station power class. This is derived from:

1. Maximum range for a single MS and single BTS.
2. Handset, cost, size, weight and battery life.

Two power classes for PCN were set 250mW and 1W peak power. This is compared with GSM900's five power classes from 800mW to 20W peak.

To derive RF performance requirements such as blocking, output RF spectrum and spurious omissions, it was necessary to define the worst case of MS-BTS coupling. This was achieved by considering the worst case of coupling between the BTS-MS, i.e. uncoordinated environment (located close to a non servicing base station antenna and located 30m away). From this a value of 65dB for the worst case coupling was agreed.

## 5.4 Overview of the PCN network

Figure 5.1 shows the main network elements within a PCN Network. The key elements are as follows:

## 152 Overview of the PCN network

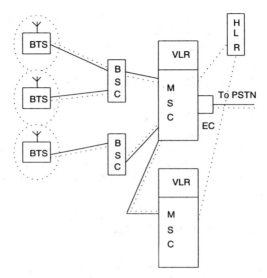

**Figure 5.1** Main elements within a PCN network

### 5.4.1 Base transceiver station (BTS)

The BTS essentially provides the radio coverage and as such provides the air interface to the customer. (The air interface will be described in more detail later.) The BTS also provides a number of 1.8GHz channels dependent upon the predicted traffic demand of the particular location. The size of the cell depends upon a number of factors, namely, terrain, traffic density, quality and local clutter. Within the PCN, system cells are expected to vary in size from a few decades of metres to several kilometres. Functions within the BTS include radio transceiver, channel coding, radio signalling, frequency hopping, paging control, access detection, measurements, encryption, radio channels.

### 5.4.2 Base station controller (BSC)

The BSC provides all the necessary local radio control features, i.e. intra-BSC handover, power control and channel allocation. The BSC also acts as a concentration site for traffic, signalling and OA & M information. As such the BSC allows flexibility in radio subsystem development strategies to provide a balance between cost and quality.

### 5.4.3 Mobile switching centre (MSC)

The MSC is the 'telephone exchange' of the network and provides call routeing and control, handover, service interworking, interconnection to other networks and echo control. Collocated with the MSC is the Visitor Location Register (VLR) which provides location update, location area information, local database and encryption key generation.

### 5.4.4 Home location register (HLR)

The HLR is the network's main database, and contains all the subscriber's information, e.g. what services the customer has subscribed to.

The specific functions of the HLR are: number translation, course location, customer profile, and charging.

## 5.5 Call set up

Consider the simplified network in Figure 5.2. 'A' is a customer on the PCN who wishes to make a call to 'C', a customer on the PSTN.

'A' is registered on the PCN. 'A' dials 'C'. The call is passed from the BTS to the MSC via the BSC. The MSC recognises that it is a call destined to the PSTN and selects an outgoing route. The route to the PSTN must include an echo canceller (EC). The PSTN routes the call to 'C'. 'C' answers and conversation commences.

The echo canceller is required due to the long delay in the radio subsystem, and as the call will be going to a two wire PSTN, customer echo would be introduced.

Now consider the situation where 'A' (customer on PCN) wishes to call 'B', also a customer on the PCN.

## 154 Planning a PCN

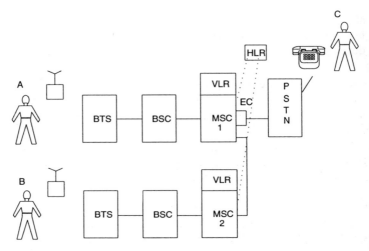

**Figure 5.2** Simplified PCN network

'A' dials 'B'. The call is passed to the MSC via the BSC. The MSC recognises that it is a call destined to B. The PCN sends a query to the HLR to find the location of B. The MSC is advised that B is registered on VLR at MSC 2 and is given B's routeing number. MSC 1 routes the call to MSC 2, MSC 2 VLR identifies B as being a particular location area. MSC 2 pages that location area. B answers, conversation begins.

## 5.6 Planning a PCN

When planning a PCN, it is necessary to consider a number of areas, including the services to be offered. The definition of services will determine the selection of infrastructure suppliers, signalling systems to be used, as well as allowing operators to develop traffic plans.

### 5.6.1 Physical network realisation

The development of the physical network implementation requires answers to:

1. What geographical coverage is required?
2. What is the traffic density per area?
3. What is the guaranteed service level, i.e. at street level or in buildings?

With this information, it is possible to plan the radio network and deploy BTS with sufficient transceivers to service demand. This then provides the customer interface.

Next the BSCs must be deployed to control the radio subsystem. Their function is to control the BTS. The location of the BSC is based upon a number of considerations:

1. How many BTSs does the BSC control?
2. What is the traffic handling capacity of the BSC?
3. Are there any other limiting factors?

Given the above, the BSC allows the network operator to optimise the BSC location in order to reduce local transmission costs.

### 5.6.1.1 *BSC support of BTSs*

The number of BTSs supported by a BSC depends upon several considerations: traffic, type of BTS and BSC, local area cost determination. The answer can range between 1 and 128 BTSs, or even higher. However, it is important to decide, when planning the network, which BTSs will be parented on to which BSC. There are three main means of connecting BTSs to BSCs: star, ring and a combination of the two.

### Typical star configuration

Figure 5.3 shows a star configuration. This is simple to implement and, subject to traffic, consists of a minimum of one 2Mbit/s path between each BTS and BSC. The main advantage of this type of configuration is that it is easy to augment routes. Growth at one particular site has no particular impact on growth at other sites.

However, the main disadvantages are:

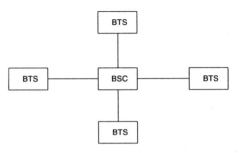

**Figure 5.3** Star configuration of BTSs

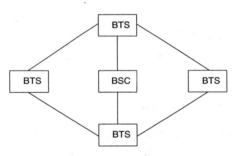

**Figure 5.4** Ring configuration of BTSs

1. Cost. It may not be possible to achieve line of sight between each BTS and the BSC and therefore it would be necessary to rent capacity, which is more expensive.
2. Reliability. Only one route exists between each BTS and the BSC which means that if we lose the link, we lose the BTS.

**Ring configuration**

Figure 5.4 shows a ring configuration. This has the main advantage in that each BTS has two routes to the BSC, thus giving an increase in reliability. However, as demand increases, unsuspected hot spots may develop.

Personal communication networks 157

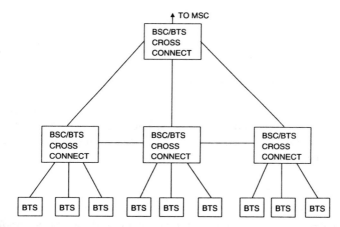

**Figure 5.5** Star BTSs within a ring of cross connects and BSCs

### Ring — star combination

To produce a general method of linking BTS to BSC, a mixture of the star and ring methods should be used, as in Figure 5.5. To allow control of capacity, it has been proposed that Digital Cross Connect (DCX) equipment is used to re-route traffic when necessary.

5.6.1.2 *BSC — MSC connection*

The connection of BSCs to MSC presents interesting opportunities for network optimisation.

As one of the functions of the BSC is to provide concentration of traffic, it can be envisaged that the utilisation of possible hubbing methods in linking BTS to BSC will ensure that traffic volumes between the BSCs and MSCs will be very high.

Given the high volume of traffic and the distances involved, protected high capacity microwave links or leased capacity will be used. However, given the economics involved, it can be expected that route diversity will need to be utilised.

## 5.6.2 Cost savings

The voice encoder used in the PCN operates at a data rate of 13kbit/s. Therefore prior to interworking with the ISDN/PSTN, the digitised voice signal must be converted to 64kbit/s A law. This is achieved via a transcoder.

The PCN architecture allows the network operator the option of positioning the transcoder at the BTS, BSC or even at the MSC. If one positions the transcoder at the MSC, this enables traffic to be carried through the radio subsystem on 16kbit/s channels.

## 5.6.3 Trunk network

The Trunk network is utilised for carrying inter-MSC traffic and for interconnecting the PCN to other networks, e.g. the national PSTN/ISDN. The interconnection of MSCs, given the distances involved, will be achieved by using high capacity bearers.

Routeing schemes are put in place to ensure that costs are optimised whilst maintaining the required quality of service.

### 5.6.3.1 *Location of MSC*

The location of the MSC is dependent upon the regulatory regime under which the PCN is being operated. For example, some countries might consider the MSC to be an adjunct to the ISDN/PSTN whereas in the UK, PCN operators are seen as independent networks. For this reason MSCs are situated in locations which achieve both on-net and off-net traffic routeing optimisation.

### 5.6.3.2 *BTS – BSC – MSC connection optimisation*

Notwithstanding the connection methods suggested above, it is necessary to ensure that location updating and handovers are optimised. This is achieved by parenting, and re-parenting BTS on BSC and BSC on MSC based upon predicted or measured traffic flows. For example, one would not have a location area boundary through the centre of Oxford Street or Times Square, nor would one have 2 BTSs in these areas parented on different BSCs.

### 5.6.3.3 *MSC – MSC routes*

Standard telecommunication practices are used to develop routeing rules for inter-MSC routes and to ensure a reasonable grade of service.

## 5.6.4 Numbering

One of the important considerations when developing the network routeing strategy for both customer traffic and signalling information, is to develop a detailed numbering plan.

Numbering plays an important part in PCNs, and the PCN companies are pressing for the introduction of a ten digit numbering scheme which will ensure that sufficient numbers are available, for both the proposed and future telecommunications services.

There are various types of numbers used within a PCN, as described in the following sections.

### 5.6.4.1 *International mobile subscriber identity (IMSI)*

The IMSI is associated with the PCN customer by being utilised on the Subscriber Identity Module (SIM). This is the plug card which, when combined with the handset, forms the mobile station.

The IMSI is the number which the network uses to identify the mobile.

### 5.6.4.2 *Mobile station ISDN number (MSISDN)*

The MSISDN is the 'telephone number' of the customer. This number conforms to ITU-T Recommendation E.164 (E.163), e.g. +44 71 492 2426 for international and 0171 492 2426 for national.

The HLR provides a mapping between the MSISDN and IMSI.

#### 5.6.4.3 *Roaming number*

This is used when it is necessary to route a call to a customer within a different service area. For instance where the customer has moved to another network, the HLR would provide a roaming number to allow the network to route the call to the customer.

#### 5.6.4.4 *Global title*

This is a number conforming to ITU-T Recommendation E.214 and is utilised to route MAP SCCP information to appropriate nodes.

#### 5.6.4.5 *Point codes*

This is used within C7 signalling networks. Point codes are used to allow signalling information to be routed through the signalling network.

## 5.7 Mobility

The philosophy behind the PCN service is 'Phones for people not places'. One must assume that the customer will move around with his telephone, unlike the fixed network where the telephone is always at the end of the line.

If customers can move, the network must always have information concerning the location of the customer to enable calls to be delivered. Furthermore the customer must also be able to change location when engaged in a call.

For the purpose of mobility, the network can be considered as depicted in Figure 5.6. The network is split into a number of location areas (LA1 to LA8).

The size of these areas is dependent upon the balance of signalling load; if the location area is small then the amount of location updating will be high. However, if the location area is large then the number of cells that have to be paged over to find the mobile will be high. RSS is the radio subsystem and incorporates BTS and BCS.

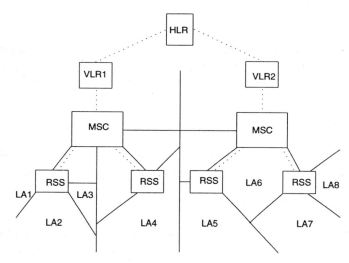

**Figure 5.6** PCN network representation to illustrate mobility

## 5.7.1 Location updating

Each PCN cell has a Broadcast Common Control Channel (BCCH) on which the location area is broadcast. The mobile monitors the BCCH channel which it can receive. When the mobile receives a new location area identification, it will send a location update request. This information updates the information stored on the VLR. If the MS enters a new VLR area, then the VLR informs the HLR. This is illustrated in Figure 5.6. Note:

1. If the mobile passes from LA2 to LA3, VLR1 information is updated.
2. If mobile passes from LA4 to LA5, VLR2 is advised, which then advises the HLR.

Once the location of the mobile has been determined, the call needs to be routed to the customer. Consider a call coming from the PSTN. The call is received at the gateway MSC which interrogates the HLR to find the MS. The HLR advises which VLR the mobile is registered on. The call is routed to the appropriate MSC. The VLR advises which LA the mobile is in and the MSC pages for the MS across the location area. The MS responds, and the call is answered and connected.

## 5.7.2 Handover

As already mentioned, people with mobile phones might want to move around when engaged in a call. It is therefore necessary for the network to maintain the call even if the customer moves between cells.

Within DCS1800 there are 4 types of handover specified:

1. Intra-BTS, a handover from one channel on a BTS to another on the same BTS.
2. Inter-BTS, Intra-BSC, a handover from a channel on one BTS to a channel on another BTS with both BTSs parented on the same BSC.
3. Inter-BSC, Intra-MSC, a handover between BTSs on different BSCs which are parented on the same MSC.
4. Inter-MSC, a handover between BTSs on different BSCs which are parented on different MSCs.

The network overhead in terms of signalling messages increases as the type of handover moves from (1) to (4). With this in mind, inter switch and inter BSC handovers should be kept to a minimum. Thus the network must be planned to achieve this.

Handover can be initiated by either the mobile or the network. The mobile will request handover as a result of radio channel measurements indicating low signal or excessive interference. The network will initiate handover for reasons such as:

1. Mobile causing interference.
2. Network management action.
3. For maintenance purposes.

## 5.8 Radio channel coding

The PCN air interface utilises TDMA techniques to conserve bandwidth. A TDMA frame consists of eight timeslots of 577ns duration. Within the PCN system, each RF channel provides eight physical channels. These physical channels can be classified into either traffic channels or control channels.

Some systems, notably in the USA, employ CDMA (Wong, 1995).

### 5.8.1 Traffic channels

These channels are intended to carry speech or data in two forms:

1. Speech channels. Full rate speech channels are defined and the algorithm is specified. However, the half rate algorithm is at present being specified.
2. Data channels. The following data rates are specified: 300, 1200, 1200/75, 2400, 4800, 9600 bits per second.

### 5.8.2 Control channels

The control channels carry signalling and synchronisation data between the handset and the network. Three types of control channels are specified, as in the following sections.

#### 5.8.2.1 *Broadcast channel*

This channel is used to broadcast information to the mobile station; the channel is downlink only. There are three broadcast channels defined:

1. Frequency Correction Channel (FCCH), for mobile station frequency correction.
2. Synchronisation Channel (SCH), for frame synchronisation of the mobile and BTS identification.

3. Broadcast Control Channel (BCCH), for the broadcast of general information on an individual basis.

#### 5.8.2.2 *Common Control Channel*

The Common Control Channel (CCCH) is used during the establishment of a connection before dedicated control channel is assigned. There are three downlink-only channels which are used for paging, access grant and cell broadcast.

The single uplink-only channel is used for random access attempts.

#### 5.8.2.3 *Dedicated Control Channel*

The Dedicated Control Channel is split into three types:

1. Standalone Dedicated Control Channel (SDCCH) carrying signalling information following mobile to network connection establishment and channel assignment.
2. Slow Associated Control Channel (SACCH) is always associated with a traffic channel or a SDCCH and maps on to the same physical channel. The SACCH carries general information from the mobile to the network such as details of current and neighbouring cell signal strengths.
3. Fast Associated Control Channel (FACCH). This channel carries the same signalling data as the SDCCH. A FACCH is assigned when a SDCCH has not been assigned and obtains access to the physical resource by 'stealing' frames from the traffic channel with which it has been assigned.

## 5.8.3 Speech coding

The PCN system uses the regular excited linear predictive coder (PRE-LPC) with long term pitch prediction. The code is block based, with blocks of 20ms, and the net bit rate is 13kbit/s. Thus one frame consists of 260kbit/s.

Following subjective tests, 182 of the 260 bits have been found to be more sensitive to bit errors and these have been called 'Class 1'

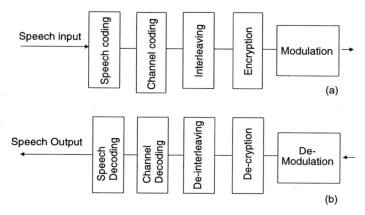

**Figure 5.7** PCN speech coding and decoding process: (a) coding; (b) decoding

bits whilst the remaining 78 bits, being robust to error, are called 'Class 2' bits.

The PCN speech coding and decoding processes are indicated in Figure 5.7. The speech coder produces the 260 bits which make up one 20ms frame. The bits are delivered to the channel coder in descending order of importance. The first 50 Class 1 bits have 3 party bits added, generated by a cyclic redundancy code. The purpose of the party bits is to detect errors.

The 182 Class 1 bits and the 3 check bits are then reordered into even bits at the start, check bits in the middle, and odd bits at the end. 4 zero tail bits are then added to give (182 + 3 + 4) = 189 bits.

The 189 bits are encoded using a convolutional code. This produces an output of 378 bits.

The 78 Class 2 bits are added giving 378 + 78 = 456. This gives a gross coded bit rate of 22.8kbit/s.

As errors in mobile radio systems are predominantly due to fading, it is possible in practice to have fades lasting a whole TDMA burst. The 456 coded bits are reordered and interleaved over 8 TDMA frames. During the interleaving process, bits are stolen for signalling purposes, i.e. the FACCH data.

Encryption is a very important feature of a mobile telecommunications service as it enables the confidentiality of conversations to be maintained. The PCN system uses inherent encryption by the authentication key Ki which is stored with the IMSI on the PCN SIM card. This information is also stored in the HLR.

### 5.8.4 Echo control

The speech channel coding introduces a one way delay in the order of 90ms. In the case of on-net calls, as separate go and return channels are used, the delay does not cause any particular problems. However, in the case of calls between the PCN and PSTN, calls will mature on analogue telephones. This means that a 2 to 4 wire hybrid will be encountered, which will introduce echo. This echo, with a delay of 90ms, could cause annoyance to customers. Therefore, it is necessary to include an echo canceller at the point of interconnection with the PSTN to cancel the echo.

## 5.9 PCN base station design

### 5.9.1 Development from GSM 900

PCN base stations are logically similar to those used for GSM 900 systems, with a number of physical enhancements to tailor them to the unique requirements of smaller cells. The main differences between GSM 900 and PCN 1800 base stations are:

1.  RF channels at 1800MHz instead of 900MHz.
2.  Smaller physical equipment size.
3.  Design optimised around large numbers of small cells, each carrying less traffic than a GSM 900 cell.

The limited number of GSM 900 technical specifications that had to be changed for 1800MHz, were released by ETSI in the DCS1800 specification in January 1991.

Personal communication networks 167

## 5.9.2 Base station structure

The Base Station Subsystem (BSS) is the part of the PCN network which connects the Mobile Station (MS) to the switch (MSC) where calls are routed to the network. Essentially, the BSS acts as a transparent bearer between the two, so that most of the enhanced services offered by PCN are provided by the MSC and not the BSS. However the 1800MHz PCN BSS is probably the most advanced cellular bearer available, and allows greater quality and quantity of traffic to be carried to the MSC than any alternative network could achieve.

The BSS consists of two network elements: the Base Station Controller (BSC), and the Base Transceiver Station (BTS) as in Figure 5.8. The BTS and BSC communicate with each other over an internal 'A-bis' interface which is represented physically by either a 64kbit/s data link or (more commonly) a 2Mbit/s data link. The connection to the MSC is via the 'A' interface, which will normally be a 2Mbit/s datalink.

In practice, a number of BTSs will be connected to each BSC. The exact number will be limited by a number of factors:

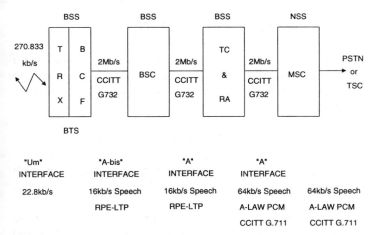

**Figure 5.8** BSS interfaces

## 168 PCN base station design

1. The number of BTS ports available on the BSC.
2. The amount of traffic generated by the BTSs (the maximum that a BSC can handle is typically around 500 Erlangs).
3. The number of transceivers (TRX) used at each BTS. Typically, a BSC cannot handle more than 100 TRX altogether.
4. The physical distance from the BTSs to the BSC. If microwave links are used, it can prove difficult to find line of sight paths to more than 20 BTSs, or even less in hilly rural areas.

To the rest of the network, the BSS looks like a single entity. The functions between the two units is allocated as:

1. Base Transceiver Station functions: interface to Mobile Stations at 1800MHz; encryption; channel coding and interleaving; frequency hopping execution; random access detection; provision of control channels; transcoding/rate adaptation; uplink measurements, including timing advance; observation and reporting of idle channels.
2. Base Station Controller functions: channel allocation; radio channel management; frequency hopping management; access grant; processing of measurements; power control; handover; paging.

Traffic channels are set up between the Mobile Station and the MSC. Speech from the handset is encoded digitally using an efficient algorithm ('RPE-LTP') which reduces the analogue speech to a 16kbit/s datastream. This 16kbit/s undergoes coding to improve the error protection, and is carried through the network as follows:

1. MS to BTS (the Um interface). Each voice channel from a mobile consists of a TDMA timeslot with a gross bit rate of 22.8kbit/s. On each RF carrier at the BTS, 8 TDMA timeslots are available.
2. BTS to BSC (the A-bis interface). Each voice channel consists of a 16kbit/s data stream. Typically, this will be carried on a 2Mbit/s datalink, with four 16kbit/s voice circuits being multiplexed onto each 64kbit/s slot.

Personal communication networks 169

3. BSC to MSC (the 'A' interface). Each voice channel will have been 'transcoded' from 16kbit/s to 64kbit/s A-Law PCM for connection to the MSC and onwards to the network. This transcoding will normally take place at the MSC end of the BSC to MSC link.

### 5.9.3 Radio frequency characteristics of PCN

The PCN BTS transmits and receives in the 1800MHz band. In the UK the following frequency allocations have been licensed:

1. Sub Band A. Ch 563 to 636. BTS Receive 1720.4MHz to 1735.0MHz. BTS Transmit 1815.4MHz to 1830.0MHz.
2. Sub Band B. Ch 671 to 744. BTS Receive 1742.0MHz to 1756.6MHz. BTS Transmit 1837.0MHz to 1851.6MHz.
3. Sub Band C. Ch 811 to 884. BTS Receive 1770.0MHz to 1784.6MHz. BTS Transmit 1865.0MHz to 1879.6MHz.

The channel spacing is 200kHz, and the transmitted signals are approximately 120kHz wide. The modulation method is GMSK with a BT product of 0.3 (i.e. the baseband data is filtered in a Gaussian filter with a bandwidth of 0.3 times the bit rate). This allows baseband data to be transmitted at 270.833kbit/s using only 120kHz of spectrum. The specifications for the receiver dynamic range, transmitter spurious emissions and intermodulation products are particularly stringent to ensure satisfactory operation in a dense traffic environment.

The highest power class for PCN Base Stations is 20W, measured at the output of the transmitter. However, before this 20W reaches the antenna, it must pass through a power combining network which couples all the transceivers in the BTS to the same antenna connector. The choice of the correct combining method is crucial to avoid unnecessary losses.

Two methods exist: hybrid combining, and tuned cavity combining. Hybrids have the advantage of small size, low cost, and broadband operation, but the losses incurred become excessive when more than 4 transmitters are to be combined. Tuned cavity filters have the advantage of low loss, regardless of the number of transmitters to be combined, but they are large, expensive, and require some method of remote retuning. Generally, hybrids will tend to dominate because PCN networks will probably have no more than 4 transceivers at the majority of sites, and the slightly higher losses will be offset by the fact that retuning is not necessary.

### 5.9.3.1 *Diversity*

PCN base stations usually give receiver space diversity. The method of combining is different from vendor to vendor, but will at least require separate receivers and two separate receive antennas. The gain on the uplink is typically 4dB or so. There is no downlink gain.

### 5.9.3.2 *Frequency hopping*

Frequency hopping can be implemented on PCN base stations. On each successive frame, the MS and BTS can change frequency giving a maximum of 216 hops per second. This gives two advantages: the exposure to interference on any one frequency is reduced, and the possibility of being in a fading null is eliminated.

The benefits of frequency hopping are mainly noticed on the downlink, as the uplink will normally have space diversity which already provides a similar effect. Frequency hopping on its own may improve the path by 3dB, but if space diversity is already in use the net improvement may be as low as 1dB.

There are two different methods of implementing frequency hopping at the base station. The most straightforward is to provide a separate transceiver for each frequency in the hopping sequence (6 is about the minimum to obtain any worthwhile benefit). Then each user's voice frames will be sent successively on different transceivers. The disadvantage of this method is that in the PCN environment only 2 or 3 transceivers are typically necessary to support a cell's traffic, and to provide 6 represents a major investment when applied across the whole network.

PCN networks will tend to use a more difficult approach, which is to actually switch the frequency of the transceiver between successive timeslots. This fast switching speed (less than 30μs) is difficult to achieve, but allows hopping to be implemented with the 2 or 3 transceivers in a typical BTS.

#### 5.9.3.3 *Discontinuous transmission*

To reduce the effects of interference, PCN mobiles and base stations usually support discontinuous transmission (DTX). This involves turning off the radio transmitter when speech is not present (which may be up to 50% of the time), using a Voice Activity Detection (VAD) algorithm in the speech coder. 'Comfort noise' is inserted by the receiving end to mask the silences that arise. This reduction in interference is particularly important in the PCN environment of small cells and high traffic density. It also has the desirable advantage of increasing the battery of the MS.

#### 5.9.3.4 *Discontinuous reception*

The MS need not have its receiver continuously turned on, provided it wakes up periodically to look for paging requests or short messages. This 'sleep mode' can improve the MS battery life quite significantly.

#### 5.9.3.5 *Dynamic power control*

A further technique for reducing interference is to allow the mobile station to use only just enough RF power to maintain acceptable quality. The power radiated by the MS can be reduced by command from the BS in 13 steps of 2dB each. The reduction is one step at a time. The decision to reduce the MS power is based on measurements of the signal level (RXLEV) and bit error rate (RXQUAL) which are made by the MS and reported to the BS.

#### 5.9.3.6 *Multipath equalisation*

Very high bit rates are used over the air interface between the handset and the base station. When a signal is reflected from a building, it may arrive at the receiver slightly after the direct wave, and cause interference. The time delay may be small (up to 10µs or so) but the bit period of the baseband data is also very short (3.7µs), so multipath equalisation is needed to demodulate the signal properly. This is achieved by transmitting a fixed bit pattern (the Training Sequence) in the middle of each time slot. Knowing this sequence, the receiver can estimate the transfer function in the time domain between the transmitter and receiver using a correlation process implemented in a Viterbi equaliser.

### 5.9.3.7 *Handover*

Handovers are used to maintain the link to the MS as it moves from cell to cell. The algorithms used in PCN networks are similar to GSM, but are tailored to the small cell environment. The two basic methods used are:

1. The 'Most Suitable Connection' method, where the mobile is always connected to the BS which provides the best path budget.
2. The 'Minimum Acceptable Performance' method where the MS is only handed over if the quality of the link drops below a certain threshold.

## 5.10 Microcells

The term microcell is often used to describe very small cells, particularly as used in PCN networks. The definition of a microcell is open to interpretation, but will generally apply to any of the following types of cell:

1. Cells with antennas positioned below rooftop or clutter height.
2. Cells with antennas inside a building (sometimes referred to as a picocell).
3. Cells covering a specific, restricted area with a typical range of less than 400m.

There are a number of significant problems associated with microcells which PCN manufacturers and operators have taken care to minimise:

1. Range. Because of the low range, large numbers of microcells are needed to provide area coverage.
2. Handover. Handover between microcells can be difficult to achieve, due to the short elapsed time required to move from one to another. Also, problems arise due to the 'street corner effect', where turning a corner may take the MS out of one cell into another with almost no overlap in which a handover can be set up.
3. Co-channel interference. Placing a large number of microcells in one area may ultimately lead to difficulties with frequency re-use. It becomes much more difficult to predict the carrier to interference ratio when the coverage is dominated by buildings rather than distance, and hence the maximum microcell density is not easy to forecast.
4. Network loading. The increased number of handovers inherent in the microcell environment lead to a greater requirement for signalling capacity in the network.

Solutions to these problems exist, and are being developed by PCN operators and manufacturers. Only when microcells are widely implemented will the advantages of PCN be fully apparent.

# 5.11 References

Avery, J. (1995) Standard serves in-building microcellular PCS, *Micorwaves & RF*, May.

Balston, D.M. (1989) Pan-European cellular radio: or 1991 and all that, *Electronics and Communications Eng. J.*, January/February.

Chia, S.T.S. (1995) Radio and system design for a dense urban personal communication network, *Electronics & Communication Engineering Journal*, August.

Cox, D.C. (1987) Universal digital portable radio communications, *Proc. IEEE*, April.

# References

Gaskell, P.S. (1991) Elaboration of the DGS1800 standard for PCN service. IEE Colloquium *GSM and PCN Enhanced Mobile Services*, IEE Digest no. 1991/023.

Goldberg, L. (1995) PCS: Technology with fractured standards, *Electronic Design*, 6 February.

Hadden, A.D. (1994) The UK PCN story, *Mobile and Cellular*, March.

HMSO (1989) *Phones on The Move*, UK Department of Industry.

Jakes, W.C.Jr. (Ed.) *Microwave Mobile Communications*, Wiley-Interscience.

Knight, P. (1991) Network planning using GSM and GSM based standards. IEE Colloquium *GSM and PCN Enhanced Mobile Services*, IEE Digest no. 1991/023.

Lindell, F., Swerup, J. and Uddenfeldt, J. (1987) Digital cellular radio for the future, *Ericsson Review*, (3).

Macaris, R.C.V. (1991) *Personal and Mobile Radio Systems*, Peter Peregrinus Ltd.

Marnick, P.J. (1991) Transmission network issues involved in the introduction of a PCN. IEE Coloquium *GSM and PCN Enhanced Mobile Services*, IEE Digest no. 1991/023.

Martin, P.C. *Networks and Telecommunications Design and Operation*, Wiley.

Mischa, S. (1988) *Telecommunications Networks Protocols, Modelling and Analysis*, Addison-Wesley.

Nielsen, M. (1995) PCS in Europe — some early lessons, *Mobile Europe*, May.

Schneiderman, R. (1995) Local loops emerge as the next big wireless market, *Microwaves & RF*, January.

Struthers Watson, K.J. (1994) PCS and the world market, *Cellular & Mobile International*, July/August.

Vincent, G. (1990) Personal communication: the dream and the reality, *IEE Review*, **36**, (8), September.

Warwick, M. (1994) PCS: first find, *Communications International*, October.

William, C.Y.L. (1986) *Mobile Communications Design Fundamentals*, Sams.

Wong, P. (1995) Clash of the Titans, *Mobile Europe*, April.

# 6. Communication satellite systems

## 6.1 Background

In 1883, Konstantin E Tsiolkovsky, a Russian schoolmaster explained the principles of rocket flight in space. Twelve years later, he mentioned the possibilities of artificial satellites circling the earth outside its atmosphere. Half a century later, Arthur C Clarke wrote of orbiting radio relay stations (Clarke, 1945) and identified many of the advantages which satellite communication would have over terrestrial systems for long distance communication and broadcasting. Above all he pointed to the unique value of geostationary satellites.

Sputnik 1, the first man-made satellite, was put into low orbit (227km × 941km × 65.1°) in October 1957 and stimulated many other tests and demonstrations of practical applications for artificial satellites in the following years, above all for telecommunications. By 1962 the Telstar 1 (Dickieson, 1963) and Relay 1 (NASA, 1968) satellites had demonstrated long distance telephone links between fixed earth stations. By 1964 an experimental satellite, Syncom III, had been placed in an accurate geostationary orbit. In the same year INTELSAT, an international consortium having the objective of setting up a global satellite network for fixed telecommunications, came into being. By July 1965 the 'Early Bird' satellite, taken over by INTELSAT and later to be renamed INTELSAT I F1, was in operation, relaying telephone calls between Europe and North America.

Much has happened since 1965 (Comparetto and Hulkower, 1995; Libbenga, 1995; Mann, 1995; Mobile, 1994; White, 1995). Satellites provide a major medium for linking together terrestrial telecommunications networks, province by province and country by country, using earth stations at fixed locations, typically with high gain antennas. INTELSAT has developed very extensively (Hall and Moss, 1978; Sachdev, 1990). INTERSPUTNIK became a second global system.

Several regional systems, such as EUTELSAT, PALAPA and ARABSAT and many national networks are now in operation.

These systems have also become an important medium for the distribution of television programme material, to terrestrial broadcasting stations and more recently direct to home. In direct to the home television these satellites are competing, not only with terrestrial radio and cable broadcasting but also with newly emerging high power broadcasting satellites. And a substantial market has developed in the supply of satellite relay facilities for linking together networks of small, low cost earth stations on the users' premises.

There were early experiments with mobile earth stations as part of the NASA Applications Technology Satellite programme in the second half of the 1960s. The MARISAT system was set up by the Communications Satellite Corporation, starting in 1976, to provide telephone and data services by satellite to merchant ships. The internationally owned INMARSAT consortium took over that function in 1982. More recently there has been widespread interest in satellite communication for airliners and for various kinds of mobile station on land; INMARSAT and a number of national systems are developing new facilities to meet these needs, often coupled with position finding aids.

## 6.2 International regulations

### 6.2.1 Frequency bands

Of the bands allocated for satellite links between fixed earth stations (ITU, 1990a), the fixed-satellite service, those given in Table 6.1 are the ones which are already used, or are likely to be used soon, for commercial systems on a substantial scale. The following notes apply to this table:

a. These bands are to be used in accordance with a frequency and orbital slot allotment plan agreed in 1987.
b. ITU Region 2 only (North and South America). In Canada, Mexico and USA, the fixed satellite service has higher allocation status than any of the other services with which the band is shared. (ITU, 1990c.)

**Table 6.1** Frequency allocations for the fixed satellite service

| Up-links (GHz) | Down-links (GHz) |
|---|---|
| 5.85 – 6.425 | 3.625 – 4.2 |
| 6.725 – 7.025[a] | 4.5 – 4.8[a] |
| 12.75 – 13.25 | 10.7 – 10.95 and 11.2 – 11.45[a] |
| 14.0 – 14.5 | 10.95 – 11.2 and 11.45 – 11.7<br>11.7 – 12.2[b]<br>12.5 – 12.75[c] |
| 27.5 – 29.5 | 17.7 – 19.7 |
| 29.5 – 30.0[d] | 19.7 – 20.2[d] |

c. Not in ITU Region 2. This band is allocated exclusively for the fixed satellite service in many countries in ITU Regions 1 and 3.
d. These bands are allocated exclusively for the fixed satellite and mobile satellite services in most countries.

These bands are also allocated for terrestrial radio systems except where otherwise indicated, and some bands are also allocated, partly or wholly, for other kinds of space system.

There are certain other bands, relatively narrow in bandwidth, also allocated for the fixed satellite service, in particular around 2.6GHz, 3.6GHz and 6.6GHz, and there are wide bands above 30GHz, but none of these bands is used much at present.

The frequency bands allocated for links between satellites (ITU, 1990a) and mobile stations (the mobile-satellite services) on land, by sea or in the air and already used for commercial systems, are shown in Table 6.2. The following notes apply to this table:

a. Not in ITU Region 1 (Europe, Africa, Asia west of the Persian Gulf and the whole of the USSR). Use for mobile satellite

**Table 6.2** Frequency allocations for mobile satellite services

| Up-links (MHz) | Down-links (MHz) | Direction unspecified (MHz) | Limitations on use |
|---|---|---|---|
| | | 806 – 890[a] | Not aircraft |
| | | 942 – 960[b] | Not aircraft |
| 1626.5 – 1631.5 | | | Ships only [c] |
| 1631.5 – 1634.5 | 1530 – 1533 | | Not aircraft |
| 1634.5 – 1645.5 | 1533 – 1544 | | Ships only [c] |
| 1645.5 – 1646.5 | 1544 – 1545 | | Distress and safety only |
| 1646.5 – 1656.5 | 1545 – 1555 | | Aircraft only |
| 1656.5 – 1660.5 | 1555 – 1559 | | Land vehicles only |

    systems is subject to protection of terrestrial systems for which the band is also allocated. (ITU, 1990d,e.)

b.   ITU Region 3 only (Australasia and Asia, excluding USSR and countries west of the Persian Gulf). Otherwise as note a.

c.   Land vehicles may use these bands for low bit rate data systems provided that they do not interfere with use by ships. (ITU, 1990f.)

The links, called feeder links, between these satellites and stations at fixed locations on the ground are operated in bands allocated to the fixed-satellite service (Table 6.1).

The frequency bands allocated for down-links from broadcasting satellites (the broadcasting satellite service) and currently being taken into use, albeit slowly, for television are as in Table 6.3

Frequency and orbital slot assignment plans have been drawn up for these bands (ITU, 1990g). The feeder links which carry programme signals up to these satellites may be assigned frequencies in

**Table 6.3** Frequency bands allocated for down-links from broadcasting satellites

| Region | Frequency band |
|---|---|
| ITU region 1 (Europe Africa Asia west of the Persian Gulf and the whole of the USSR) | 11.7 – 12.5 GHz |
| ITU region 2 (North and South America) | 12.2 – 12.7 GHz |
| ITU region 3 (Australasia and Asia excluding USSR and countries west of the Persian Gulf) | 11.7 – 12.2 GHz |

**Table 6.4** Frequency bands allocated for feeder links used in broadcast satellite services

| Region | Frequency band |
|---|---|
| Regions 1 and 3 | 14.5 – 14.8 GHz and 17.3 – 18.1 GHz |
| Region 2 | 17.3 – 17.8 GHz |

bands allocated to the fixed satellite service for up-links. However, other frequency bands, not to be used for fixed satellite systems, have been set aside especially for these feeder links and assignment plans have been drawn up. The plans are in the bands given in Table 6.4 (ITU, 1990h).

Various other bands have been allocated for satellite broadcasting (down-links), in particular around 42GHz and 85GHz, but there is little sign at present of these allocations being taken into use.

Finally, frequency bands have been allocated (ITU, 1990a) for assignments to inter-satellite links, used for any kind of satellite system. The main bands below 100GHz are: 22.55GHz–23.55GHz; 32.0GHz–33.0GHz; 54.25GHz–58.2GHz; 59GHz–64GHz.

## 6.2.2 Constraints on frequency assignments

Satellite systems share the frequency spectrum with other satellite systems and, in most frequency bands, with terrestrial radio systems also. Consequently interference arises. In order to keep the interference down to acceptable levels, constraints on the characteristics of systems which determine their liability to cause and suffer interference have been agreed. Also, assignments are registered internationally; a new user of a frequency defers to a registered established user under the jurisdiction of another country where there is no international agreement to the contrary. It is assumed that interference problems between systems falling within the jurisdiction of the same country will be resolved by the government of that country. The constraints and agreements vary from service to service and from band to band; the more important provisions are as follows.

In the frequency bands at 12GHz in which frequency assignment plans have been agreed for satellite broadcasting (ITU, 1990g) and in the bands at 14.5GHz and18 GHz where there are corresponding feeder link assignment plans (ITU, 1990h), limits have been agreed for all of the satellite and earth station characteristics and all of the emission and orbital parameters which significantly affect the level of interference that these systems can cause to one another. With very minor exceptions, the radio stations of other services which also have allocations in these bands are not permitted to cause significant interference to satellite broadcasting, and they must accept any interference which they receive from authorised satellite broadcasting.

In the frequency bands at 4.5, 7.0, 10.8, 11.3 and 13GHz, which were used in drawing up the frequency and orbital slot allotment plan which provides for one fixed satellite system per country (ITU, 1990b), constraints were placed on equipment characteristics and emission and orbital parameters in more flexible ways than were used for satellite broadcasting at 12GHz. There are also procedures to be used when systems are set up, using these allotments, to ensure that these systems do not interfere with one another. However, these frequency bands are also allocated for terrestrial radio services and the measures outlined below to limit interference between terrestrial and satellite systems in bands which have not been planned for the satellite service apply in the allotment plan bands also.

These two sets of frequency plans for satellite systems are limited to geostationary satellites. Most communication satellites operating in other bands are also geostationary, but a few use other orbits. Satellites in non-geostationary equatorial orbits would present a persistent interference hazard for geostationary satellites. Non-geostationary satellites in orbits inclined to the equator, which pass through the equatorial plane twice in every orbit, could cause and suffer intermittent interference. Since it is agreed that the geostationary satellite orbit is the most important, especially for the fixed satellite service, there is a regulation (RR 2613) that requires systems using non-geostationary satellites using fixed satellite allocations must cease operation whenever this is necessary to prevent interference to a geostationary satellite system (ITU, 1990i).

There are also agreed procedures, implemented under the supervision of the ITU, to ensure that the owners of proposed new geostationary satellite systems, and the national administrations under whose jurisdiction they fall, will meet with corresponding representatives of established geostationary systems belonging to other countries, to make sure that interference problems will not arise; this is called frequency co-ordination (ITU, 1990j). When frequency co-ordination has been successfully completed and the new system has been brought into operation, frequencies assigned to the satellite can be registered in the Master International Frequency Register, maintained by the ITU. Procedures intended to deal systematically with the risk of interference between non-geostationary satellites have not yet been developed.

To keep within acceptable limits the interference which satellite transmitters cause at the receiving stations of terrestrial radio links, upper limits have been placed on the spectral power flux density (PFD) which satellites, regardless of orbit, may set up at the Earth's surface in frequency bands shared with terrestrial services (ITU, 1990k). These limits vary with frequency and with the angle of elevation at which the satellite signal reaches the ground. However, a limit of $-145$dB relative to 1 watt per $m^2$ in a sampling bandwidth of 4kHz is typical for frequency bands below 15GHz. Corresponding limits apply to the power of terrestrial transmitters operating in bands in which satellite receivers operate.

Finally there is the problem of interference between earth stations operating with satellites and the transmitting and receiving stations of terrestrial radio systems. As usual, national governments must resolve problems arising between stations under their jurisdiction. However, if a new earth station could suffer unacceptable interference from a foreign terrestrial transmitter or vice versa, the extent of the problem must be determined and solutions found where necessary before frequency assignments made to the earth station may be registered internationally (ITU, 1990l).

## 6.3 Spacecraft technology

### 6.3.1 Orbits

By far the most useful orbit for communication satellites is the geostationary satellite orbit. This is a direct equatorial orbit about 35800km above the ground, the period of which is the same as the length of the sidereal day, about 23 hours 56 minutes. A satellite in this orbit, moving in the same direction as the Earth's rotation, remains stationary as seen from points on the Earth's surface. A geostationary satellite has line of sight coverage of a great area of the Earth and, as Clarke noted in 1945, three of them suitably located around the Equator could cover almost all the Earth's surface. Figure 6.1 shows, for example, the coverage provided by satellites located at $30°W$, $150°W$ and $90°E$ longitude.

There are, however, other orbits of interest for satellite communication. The USSR, with a territory which has an exceptionally wide span in longitude at high latitudes, found 12-hour elliptical orbits inclined at about $63°$ to the equatorial plane to be preferable for its domestic ORBITA network. These satellites are operated for periods of about eight hours when they are close to their apogee, about 40000km high above Siberia, but three satellites are required to provide continuous coverage.

Satellites in geostationary and 12-hour elliptical orbits are at great distances from the surface of the Earth and the transmission loss is

## Communication satellite systems 183

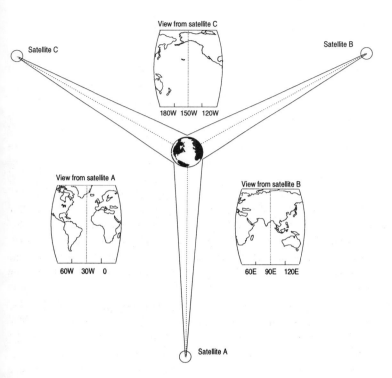

**Figure 6.1** The concept of a world-wide geostationary satellite system

very high; at 1.6GHz, for example, the free space loss is about 188dB. This loss has particularly serious consequences for satellite communication with mobile earth stations and in particular for hand portable and vehicle mounted stations, which typically have little antenna gain.

Links with satellites in low orbits would have significantly less loss; for example, for an orbit only 1500km above the ground the loss would be about 28dB less, which would make a great difference to the weight of batteries required in a pocket radio telephone. For reasons such as these, low orbits are being seriously considered for this application.

## 6.3.2 Launchers and launching

The classic procedure for launching a geostationary satellite falls into three phases. In the first phase (boost phase) a powerful two stage liquid propellant rocket (the booster) places the satellite and the necessary upstage rockets in an orbit about 200km above the ground. This orbit, called the 'parking' orbit, has a period of about 100 minutes, it is direct (that is, the satellite revolves in the direction in which the Earth rotates), it is approximately circular and the plane of the orbit is inclined to the plane of the equator unless it happens that the launching site is located on the equator. The booster will have been jettisoned before this parking orbit is attained and the second stage rocket may be detached in the parking orbit.

In the second phase, as the satellite is passing through the equatorial plane, another rocket, called the 'perigee motor', is fired to accelerate the satellite out of the parking orbit and into the elliptical 'transfer' orbit. This orbit has a period of about 11 hours, and the same inclination to the equatorial plane as the parking orbit. The perigee motor casing is then discarded. Close to its apogee and about 35800km above the ground, this transfer orbit passes through the equatorial plane.

In the third phase, at a time when the satellite is close to its apogee and about to pass through the equatorial plane, the firing of the final stage rocket, called the 'apogee motor', accelerates the satellite to a velocity of about 3kp/s and removes the inclination of its orbit. This change of velocity and direction causes the orbit to become roughly circular and equatorial and its period becomes roughly equal to one sidereal day; it will, in fact, be approximately geostationary. Low-energy thrusters incorporated in the satellite itself, typically fuelled with hydrazine, are then used to move the satellite to the point in longitude at which it is to operate and to correct any remaining errors in circularity, period or orbital inclination.

The United States Space Transportation System (Space Shuttle) enables the first phase of launching to be carried out by reusable rockets, although this facility has not been available for commercial launching since 1986. Other reusable launch and upper stage systems are under development. However, all other launch vehicles in current use are expendable.

The classic procedure described above can be varied to take advantage of the capabilities of particular launch vehicles. 'Strap-on' solid propellant rockets are usually added to the booster to increase the payload mass which it can launch. The second stage rocket may also perform the functions of the perigee motor, allowing the first and second phases of the launch procedure to be fused into one phase. A single liquid propellant rocket may combine the functions of apogee motor and perigee motor. This, in fact, is the procedure that is most commonly used today; it is illustrated in Figures 6.2 and 6.3. Despite such variants, however, the principles of the classic procedure are generally applicable for all geostationary launchings and may also be used, with appropriate changes, for launching satellites which are not to be geostationary.

Launching facilities are available from a number of providers, government and commercial, in USA, USSR, France, Japan and China. The mass which these launch vehicle systems can deliver into, for example, the geostationary satellite orbit ranges from 350kg to 4000kg, and the more powerful systems can also be used to deliver several light weight satellites into the same, or similar, orbits. A comparison between the payload capabilities of major communication satellite launchers is shown in Table 6.5.

## 6.3.3 Orbital perturbations and their correction

In time, natural forces change the orbits into which satellites are initially put. These perturbations are most significant and most thoroughly characterised for geostationary satellites.

Non-uniformity of the gravitational field of the Earth causes the period of a geostationary satellite, initially exactly equal to one sidereal day, to increase or decrease so as to accelerate the satellite towards either of two stable points, at $77°$ east longitude or $108°$ west longitude. In many parts of the orbit a significant drift in location, due to this cause, accumulates within a few weeks if not corrected.

The gravitational fields of the Sun and the Moon cause the orbital plane of a geostationary satellite, initially co-incident with the equa-

186 Spacecraft technology

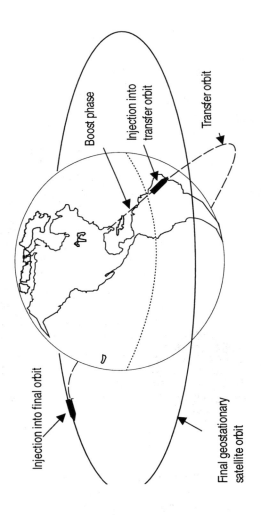

**Figure 6.2** Flight profile to the geostationary satellite orbit

Communication satellite systems 187

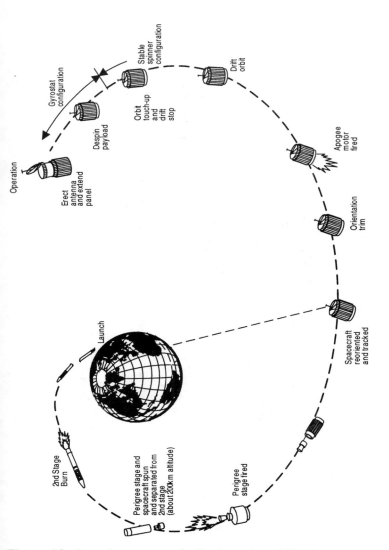

**Figure 6.3** Launch sequence of a Hughes spin stabilised geostationary communications satellite using a Delta 3910 expendable launcher

**Table 6.5** Characteristics of the major communication satellite launchers

| | Space shuttle | H-1 | Atlas 2 | Delta II | Proton 4 | Ariane 4 | Long March 3 | Titan IV |
|---|---|---|---|---|---|---|---|---|
| Country | USA | Japan | USA | USA | USSR | Europe | China | USA |
| Developer | NASA/ Rockwell International | NASDA/ Mitsubishi | General Dynamics | McDonnell Douglas | Glancosmos | ESA/CVES | CAST Ministry of Machines | Martin Marietta |
| Length × diameter (m) | 56.1 × 23.79 | 35.35 × 2.44 | 475. × 3.05 | 38.4 × 2.44 | 44.3 × 7.4 | 57.1 × 3.8 | 43.83 × 3.35 | 60 × 3.05 |
| Mass (kg) | 2040815 | 135000 | 187560 | 228100 | 700000 | 243000 | 202000 | 800000 |
| Capability | Due east launch from Cape Canaveral. Maximum payload into low earth orbit 29478kg | 350kg geostationary orbit | 6780kg to low earth orbit; 2740kg to geostationary transfer orbit | 539kg to 28.7° 185km; 3819kg to 90° 185km. Uses upper stages for higher orbits | Using 4th stage block DM 2120ks to geostationary transfer orbit | 1900kg to geostationary transfer orbit; 4600 to low earth orbit | 1400kg to 31° 200-35500km geostationary transfer orbit | 1770 to 28.6° low earth orbit. Use Centaur or ISU for geostationary transfer orbit or geostationary orbit |
| Year entering service | 1981 | 1986 | 1991 | 1991 | 1967 | 1989 | 1984 | 1989 |

torial plane, to become inclined to it. The inclination grows at about $0.86°$ per annum if not corrected, causing daily excursions of the satellite north and south of its nominal position.

The pressure of solar radiation and solar wind on a geostationary satellite cause the orbit, initially circular, to become somewhat elliptical. The effect is seasonal, a build up of ellipticity at one time of year being neutralised by a change in the contrary sense at another time of year. Ellipticity causes the satellite to seem to oscillate daily east and west of its nominal position. The effect increases with the area of cross-section of the satellite and it is big enough to be significant if the satellite is large.

Significant movement of a satellite, east or west of its nominal position, is likely to cause interference to or from another network operating in the same frequency band and using a neighbouring satellite. Drifts due to the Earth's gravitational field can be corrected throughout the life of a satellite by means of the same Hydrazine thrusters as were used for the final adjustment of the orbit during the launch procedure; the amount of fuel required for this is relatively small.

The ITU Radio Regulations require geostationary satellites using frequency bands allocated to the fixed satellite service to be maintained within $0.1°$ of their nominal orbital longitude, if an east-west perturbation would cause unacceptable interference to another system (ITU, 1990m). The constraint is eased to $0.5°$ for experimental satellites and satellites which do not use fixed satellite frequency bands (ITU, 1990n).

A further relaxation, to $1.0°$, applies to certain old satellites, launched before 1987 (ITU, 1990o). Constraints on broadcasting satellites operating at 12GHz are at least as severe, and they are unconditional (ITU, 1990g).

North-south excursions of satellites due to inclination of the orbital plane can also be corrected by the on board thrusters, but the amount of fuel that is required to maintain low inclination is quite significant; for a 10-year lifetime the mass of hydrazine required for this purpose is about 20% of the total mass of the satellite at start of life.

For this reason, bi-propellant thruster systems and electrically powered ion engines which use less payload mass are coming into use

(Hayn, et al., 1978; Free, 1980). No regulatory limit is applied to excursions north and south of the equatorial plane, but satellites with inclinations of more than $5°$ have not been regarded as geostationary for regulatory purposes.

### 6.3.4 Attitude stabilisation

Stabilisation of a communication satellite's attitude relative to the Earth is necessary, in order that the gain of directive satellite antennas may be used to use satellite power efficiently and to permit geographical re-use of spectrum. Spinning body stabilisation has been widely used in the past but three axis body stabilisation is in general use now.

In a spinning body satellite the satellite body rotates at 30rpm to 100rpm about the axis which is perpendicular to the plane of the orbit. Antennas are usually mounted on a rotatable platform which is 'de-spun' relative to the Earth and is, accordingly, stabilised in all three axes. A reference to enable the on board control system to keep the antenna platform continually pointing towards the Earth is usually obtained from infra red Earth sensors, supplemented by Sun sensors. Antenna pointing accuracy of $\pm 0.2°$ or better is obtained with such systems.

The thrusters which are used for east-west orbit adjustment are usually mounted on the rotating body of the satellite, and these thrusters accordingly must be operated in a pulsed mode, synchronised with the rotation of the body.

Body stabilised designs generally employ an internal momentum wheel with its axis perpendicular to the plane of the orbit. Control of attitude about the pitch axis is obtained by varying the speed of rotation of the wheel. Hydrazine thrusters are used occasionally to dump momentum from the wheel, so avoiding an unacceptable build up of rate of rotation. Control about the yaw and roll axes may be obtained by gimbaling the wheel or by the use of hydrazine thrusters. Figure 6.4 shows INTELSAT V, a typical large body stabilised satellite (Fuenzalida et al., 1977).

The ITU Radio Regulations require geostationary satellites in general to be capable of maintaining their antenna beams within 10%

Communication satellite systems 191

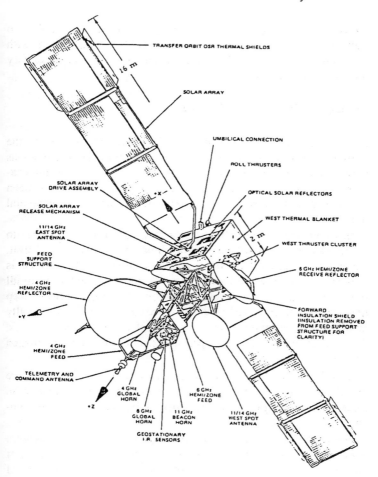

**Figure 6.4** INTELSAT V satellite configuration

of the half power beamwidth, or ±0.3°, of the nominal direction, whichever requirement is less stringent (ITU, 1990p). The frequency assignment plans for satellite broadcasting at 12GHz require beam pointing within ±0.1° (ITU, 1990g).

## 6.3.5 Electrical power supply in space

Silicon solar cells are used as the primary satellite power source in normal operation. When the Sun, for the satellite, is eclipsed by the Earth, power is maintained by nickel cadmium or nickel hydrogen secondary batteries. Full shadow eclipse for geostationary satellites occurs on 44 nights in spring and 44 nights in the autumn, the longest eclipses occurring at the equinoxes and giving 65 minutes of full shadow. Research on new types of storage cell, such as the silver hydrogen cell, seeking longer life, lower mass and higher efficiency, is in progress.

Spinning body satellites have body mounted arrays of solar cells, typically producing about 10 watts per kilogram of solar array mass. Body stabilised satellites using extendable arrays, which can be rotated so that they always face the Sun, deliver up to 23W/kg. Current research is aimed at increasing the dimensions of deployable solar panels and reducing their mass. Values are reaching 50 to 60 watts per kilogram and a few tens of kilowatts per panel. High efficiency solar cells now being developed could cut substantially the mass and size of arrays.

One disadvantage of deployed arrays is the limited amount of power available before the array can be deployed, that is, while the satellite is still in its transfer orbit.

## 6.3.6 Telemetry, tracking and command

Telemetry, tracking and command (TT&C) facilities are needed, in the launch phase and in normal operation of communication satellites, for a variety of purposes. Telemetry is used to monitor and evaluate the performance and behaviour of the satellite and to provide the ground control station with data for the diagnosis of fault conditions that may arise. Tracking facilities, used in conjunction with the command system, enable the location, velocity and orbit of the satellite to be determined, enabling the orbit and the nominal location in orbit to be set up initially, then maintained throughout the satellite's operating life. Command facilities are required for the initiation of all the manoeuvres which cannot be automated for on board control and which are required of the satellite as a vehicle and as a complicated piece of communications equipment.

The signal channels between satellite and control Earth stations required for these functions are carried by two radio subsystems. One subsystem operates at low microwave frequencies allocated to the space operations service, using low gain satellite antennas; this subsystem is used during the launch phase, when TT&C may be required, for example between the launch site and the satellite and the attitude of the satellite is not normal. The other subsystem operates in a narrow band within the frequency bands appropriate to the mission of the satellite and uses the mission satellite antennas, often of high gain; this subsystem is used during the normal operation of the satellite.

## 6.4 The communication chain

### 6.4.1 The chain in outline

Figure 6.5 represents the communication chain in satellite communication in its simplest form. At one Earth station, signals from a source modulate a carrier (M), the carrier is up-converted (U/C) to a radio frequency suitably for the up-link, amplified in a transmitter (Te) and radiated to the satellite receiving antenna. At the satellite, the received carrier is amplified (Rs), changed to the down-link frequency (F/C), amplified in a transmitter (Ts) and radiated to another Earth station; this assembly of equipment in the satellite is called a transponder. At that second Earth station the down-link carrier is received and amplified (Re), down-converted (D/C) and demodulated (D) so that the signal can be passed to its destination.

In practical systems the communication path is often duplex, the return signal channel usually being transmitted through the same satellite transponder. A carrier transmitted from one Earth station may be received at many Earth stations. There may be several or many other pairs of Earth stations passing carriers through the same satellite transponder simultaneously or in sequence, which is called multiple access. There may be other transponders in the satellite, isolated from one another by frequency separation, satellite antenna directivity or polarisation discrimination and relaying different groups of carriers.

194  The communication chain

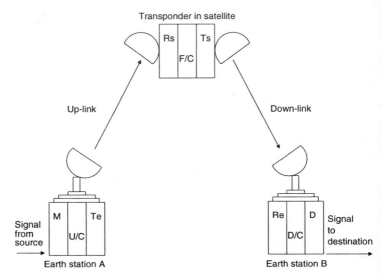

**Figure 6.5**  The basic satellite communication chain

The techniques which allow satellite communication to be used in these various ways are reviewed later in this chapter.

The optimum choice of orbit for a satellite depends primarily on the mission. The geostationary satellite orbit (GSO) has several important advantages over other orbits that might be used for communication systems. A geostationary satellite, being stationary in the sky as seen from the Earth, is constantly in sight of a fixed Earth station. The variation with time of the transmission delay is zero for an ideal geostationary satellite and it is small where good standards of satellite station keeping are maintained. Satellites in other orbits, such as those illustrated in Figure 6.6, move across the sky, so Earth station antennas, if they are directional, must track the satellite movement.

Most non-geostationary satellites periodically disappear below the horizon; if unbroken operation is required, several or many satellites must be deployed in suitable configured orbits and the communication links between Earth stations must be transferred from one satellite to another at intervals ranging from several times per day down

Communication satellite systems 195

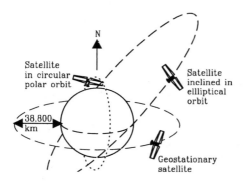

**Figure 6.6** Some typical communication satellite orbits (not strictly to scale)

to several times per hour. Furthermore the motion of these satellites, relative to the Earth stations, causes the Earth station to Earth station transmission time to vary with time, leading to synchronisation problems with high speed digital systems and Doppler shift of signal frequencies in analogue systems.

However, the GSO has two serious disadvantages, which it shares with some elliptical orbits. The free space transmission loss between a satellite and an Earth station and the transmission delay, Earth station to Earth station, are both very large. Figure 6.7 shows how these quantities vary with the altitude of the satellite above the Earth's surface (the transmission loss also varies with carrier frequency and the figure shows values for 10GHz).

Lower transmission loss for low altitude offers big advantages for some applications.

## 6.4.2 Space-Earth propagation

In the early years of satellite communication the cost per watt of radio frequency power from a satellite transmitter was very high. Technical developments have reduced that cost, but economy in the use of satellite power remains a factor that dominates system design. Thus, the power of the carrier wave that is intended to reach an Earth station receiver is determined rather precisely, to provide an adequate carrier to noise ratio for a sufficiently high proportion of the time, but no more.

The free space transmission loss from a geostationary satellite to an Earth station is very much bigger than the corresponding loss for any terrestrial line of sight path. Figure 6.7 shows that a typical value for this loss is 204dB at 10GHz; this falls to 196dB at 4GHz and rises to 214dB at 30GHz. This loss is the principal factor determining what the satellite power level will be. However, a margin of power must be provided in addition to ensure that channel performance targets are

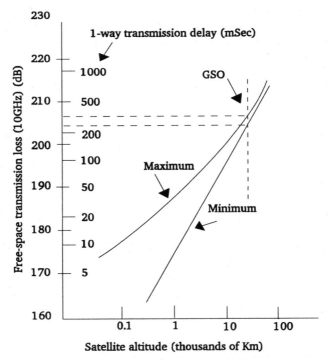

**Figure 6.7** Transmission loss and decay as a function of satellite altitude

reached. This margin provides in part for under achievement of equipment performance and operating targets and in part for propagation loss, due to absorption in the troposphere, which is additional to the free space value.

The principal cause of tropospheric loss at frequencies in use or in prospect of use for satellite communication is absorption in rain. Molecular resonance in water vapour begins to be significant as the frequency rises above 20GHz. The extent of the loss due to rain varies greatly, with frequency and climate but also with the angle of elevation of the satellite as seen from the Earth station, and Figure 6.8 can give no more than a very general indication of how these losses vary with frequency under given weather conditions.

The margin of satellite carrier power that must be allowed, to protect link performance against the effects of rain, depends significantly on the percentage of the time during which deterioration of link performance can be accepted.

In reading Figure 6.8, it may be noted that a rainfall rate of 10mm/hour can be expected in many temperate climates for 0.1% of the time and that rates of 65mm/hour are found for 0.1% of the time in tropical high rainfall areas.

Significant depolarisation of signals arises in heavy rain, and it may be necessary to use adaptive compensation for this at Earth stations in rainy climates which depend on a large measure of polarisation discrimination.

## 6.4.3 The transponders

The first INTELSAT IV satellite was launched in January 1971 but the design plan of the transponders for these satellites was highly innovative (Jilg, 1972) and it has remained a basic model for most satellites that have followed. The satellite used the 6GHz band for up-links and the 4GHz band for down-links. Figure 6.9 shows the basic elements.

Signals in the band 5.932GHz to 6.418GHz from the receive antenna were first amplified in a 6GHz tunnel diode amplifier (TDA) and then frequency translated by 2225MHz for further broadband amplification in a 4GHz TDA and a low level travelling wave tube

198  The communication chain

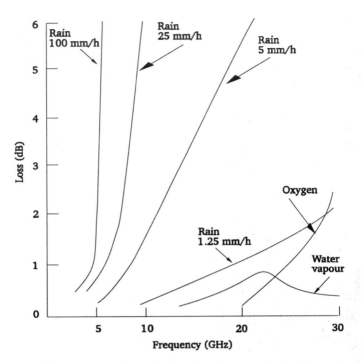

**Figure 6.8**  An approximate indication of tropospheric absorption on space-earth paths

(TWT). This part of the system had fourfold redundancy to achieve high reliability and long life. The signals were then split by a filter dividing network into 12 channels, each 36MHz wide and 40MHz spacing between centre frequencies. Each channel had its own redundant high level TWT amplifier rated at 6W single carrier saturated output power. The gain of the amplifiers for channels 1 to 8 was controllable in eight 3.6dB steps; for channels 9 to 12 there were four steps. The outputs of all of these amplifiers, after bandpass filtration to remove out of band distortion products, could be connected to one of two horn antennas covering the whole visible disc of the Earth. Alternatively, any or all of channels 1, 3, 5 and 7 could be connected by telecommand to spot beam antenna number 1 (half power beam-

Communication satellite systems 199

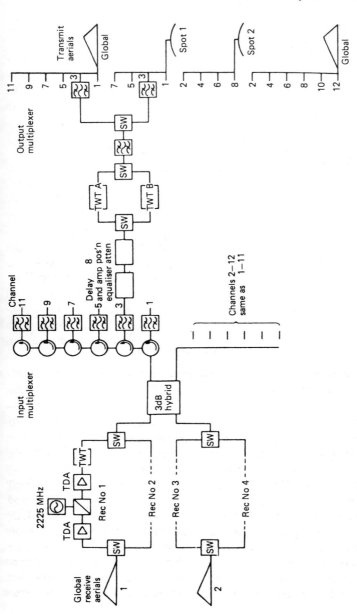

**Figure 6.9** INTELSAT IV transponder

width 4.5°) and similarly channels 2, 4, 6 and 8 had optional access to spot beam antenna number 2.

The principal characteristics of the transponder and antenna subsystems were as follows:

1. Receive system gain to noise temperature ratio (G/T) equal to −17.6dB/K.
2. Up-link power flux density at receive antenna for output TWT saturation equal to −73.7dBW/m$^2$ to −55.7dBW/m$^2$.
3. Transmit e.i.r.p. at single-carrier saturation: global beam equal to 22dBW per channel; spot beam equal to 33.7dBW channel.
4. The receive antennas were LH circularly polarised and the transmit antennas were RH circularly polarised.

By way of contrast, the EUTELSAT II design is very new, the first being launched in 1991. The 14GHz band is used for up-links and the 11GHz and 12GHz band for down-links. The system employs 'frequency re-use' by dual linear polarisation, providing up to 1GHz of usable bandwidth, some accessed by narrow band (36MHz) transponders and some by wide band (72MHz) transponders. All transponders have redundant TWT amplifiers rated at 50W single carrier saturated output.

Where the demand for satellite services is high it is economically advantageous to maximise the bandwidth that is available for signals. Many satellites, like EUTELSAT II, obtain two fold frequency re-use by dual polarisation. Many satellites are equipped to operate in more than one pair of frequency bands, typically having some transponders operating at 6GHz and 4GHz and others operating at 14GHz and 11/12GHz. In satellites serving extensive geographical areas, the transponder capacity may be enhanced by frequency re-use between non-overlapping spot beams. Thus, applying all of these frequency use techniques, the INTELSAT VI satellites have 38 transponders which use the 6 and 4 GHz bands 8 times over, plus 10 transponders which use the 14 and 11/12 GHz bands twice over.

At the sub-unit level, major changes are taking place. Solid state amplifiers have largely replaced TWTs in power amplifiers at 4GHz. There is a trend away from transponder power to bandwidth ratios appropriate to high performance Earth stations and towards more powerful transponders suitable for small antenna Earth stations. Most radically, transponders and on board signal path switching systems are becoming optimised for digital transmission systems. On multibeam satellites, in addition to long term flexibility switches which enable signal paths to be set up on a semi-permanent basis between a specified up-link beam and a specified down-link beam within a given frequency band, switch matrixes capable of reconfiguring beam to beam connections from millisecond to millisecond are now being taken into use for on-board switched time division multiple access systems (Watt, 1986). Development work is also in progress on on-board signal processing equipment which can demodulate up-linked digital signals, regenerate the digital waveform and assemble the signals in appropriate time division multiplex streams for more economical transmission back to Earth (Evans, 1986).

### 6.4.4 Satellite antennas and footprints

A circular beam $17.5°$ wide is just sufficient to cover the whole disc of the Earth visible from a geostationary satellite. Such a beam has a beam edge gain of about 16dBi and it is generally provided by a horn antenna.

For most purposes, coverage of the whole visible Earth is not necessary, in which case it is also not desirable. Antennas with higher gain which, nevertheless, cover the required service area can provide much greater transmission capacity for given transponder power than global coverage antennas. Furthermore, frequency co-ordination of such satellites with others operating in the same frequency band is made easier, and the possibility may arise for frequency re-use within the satellite network if beams do not overlap.

The basic high gain antenna is an offset fed reflector, generating an approximately circular beam. The beam edge gain of such an antenna is given approximately by Equation 6.1 where D is the half power beamwidth in degrees.

$$Gain = 41 - 20 \log_{10} D \quad dBi \tag{6.1}$$

## 202 The communication chain

However, it is so important to obtain the highest feasible gain from a satellite antenna that it is usual to adopt a more complex antenna design, producing a pattern of illumination on the Earth's surface (called the 'footprint') which matches closely the geographical area which is to be served. Thus, for example, INTELSAT V has six antennas for access to and from the transponders and two of them, optimised to generate beams from a mid-Atlantic orbital location which serve areas on both sides of the Atlantic Ocean where large amounts of traffic originate, take the form of front fed reflectors with an array of 88 carefully phased feed horns at their offset foci. (See Figure 6.10.) The corresponding feed horn arrays on INTELSAT VI antennas have 146 elements.

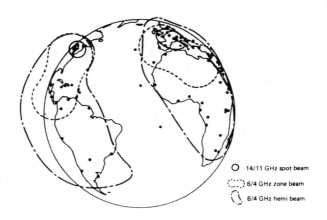

**Figure 6.10** Approximate aerial beam coverage of INTELSAT V spacecraft for the Atlantic Ocen region

## 6.4.5 Modulation techniques

Up to the present time the choice of modulation technique for use in satellite communication has been greatly influenced by the cost of carrier power reaching the receiving antenna. This cost is tending to fall, in particular as satellites with higher antenna gain come into use, but another limitation on down-link power levels is likely to remain as long as most spectrum allocated for space services is shared with terrestrial radio services which have equal allocation status.

Despite the low receiver noise levels obtainable with even low cost Earth station receivers, modulation techniques must be suitable for operation at relatively low carrier to noise ratios.

Amplitude modulation is never used for analogue signals and the use of high order phase shift and hybrid modulation for digital signals is rare. However, modulation techniques which tolerate lower pre-demodulator carrier to noise ratios (C/N) tend to need wider bandwidth for a given information capacity. Thus the modulation parameters should be optimised for each situation, to ensure that the best use is made of transponder capacity.

Frequency modulation is most commonly used for signals which are radiated in analogue form and a relatively high index of modulation is typical.

Ideally, the index of modulation, and therefore the bandwidth occupied before demodulation, and the carrier power level are chosen so that:

1. The threshold of the demodulator under clear sky conditions (that is, in the absence of signal absorption in the troposphere) will be exceeded by a few decibels (related to the required rain and implementation margins) and the necessary post demodulator signal to noise ratio (S/N) will be attained for the specified proportion of the time.
2. The power and bandwidth available from the transponder for the carrier, and any other carriers that the transponder may be relaying, will be occupied when the transponder is fully loaded.

The choice involves consideration of many of the characteristics of the satellite network, the most important of which is probably the figure of merit (G/T) of the antenna and receiver combinations of the various Earth stations involved.

Wide deviation FM is widely used for analogue television signals and frequency division multiplex (FDM) multi-channel telephone baseband aggregates. When used for single speech channels, a valuable saving in power can be obtained by suppressing the carrier when the speaker is silent. The concentration of spectral energy in the neighbourhood of the carrier frequency of an FM signal may necessitate the application of a carrier energy dispersal waveform to the baseband of a television emission or a high capacity FDM telephone emission, to meet the PFD constraint referred to earlier, if the down-link frequency allocation is shared with terrestrial radio services.

For digital signals, phase shift keying (PSK) is most commonly used, 2-phase or 4-phase. For 2-phase PSK a clear sky C/N ratio of 8.4dB, plus a small rain and implementation margin, is sufficient and for 4-phase PSK the ratio should be 3dB larger. However, some form of forward error correction (FEC) is often used, especially where the G/T of the Earth station receivers is low and this permits satisfactory operation with a significantly lower C/N ratio.

Such emissions carry all kinds of digital signals, ranging from single speech channels, through time division multiplex (TDM) multi-channel telephone aggregates and data systems of a wide range of information rates to digital television signals, although the last mentioned are usually subjected to some form of bit rate reduction video signal processing.

Wide band digital emissions may exhibit strong spectral lines under idle circuit conditions, and it may be necessary to add to the modulating signal at the transmitting Earth station a pseudo-random sequence, to be subtracted at the receiving Earth station, to disperse these lines if the down-link frequency allocation is shared with terrestrial radio services.

When the G/T of receiving Earth stations is very small, as it may be with some very small aperture terminal (VSAT) networks, it may be preferable to disperse the spectral energy of the carrier by using frequency hopping spread spectrum modulation.

## 6.4.6 Multiple access methods

Communication satellites are designed to relay several, or more usually many, signals simultaneously. In some cases there may be a separate transponder for each carrier; this is typical of broadcasting satellites and of satellites used for distributing television signals to terrestrial broadcasting stations. More usually, each transponder will relay, not one carrier, but several or many. This is called 'multiple access'. There are three basic techniques for achieving multiple access without unacceptable interference between the various signals involved.

In frequency division multiple access (FDMA) the carriers that will be relayed by a transponder are assigned carrier frequencies within the transmission band of the transponder, the frequency separation between assigned frequencies being sufficient to avoid overlap of emission spectra. The travelling wave tube (TWT) and solid state power amplifiers which are used in transponders have relatively constant gain characteristics within a certain range of drive levels, but they become non-linear, then saturate, as an upper limit is approached. Therefore the output of the transponder will contain the input carriers, amplified, plus distortion products, such as harmonics of the carriers and the products of intermodulation between them, the level of which will be high if the input carrier aggregate is powerful enough to drive the amplifier close to saturation (Westcott, 1972; Chitre and Fuenzalida, 1972).

For a transponder operating in the FDMA mode, the power level of each up-link carrier reaching the satellite must be set with two objectives. The first is to obtain at the output of the amplifier the optimum ratio between useful carrier power and noise due to the distortion products in the vicinity of the carriers. This involves backing-off the aggregate input level from the point where the amplifier would be driven to maximum total output, in order to obtain a larger reduction in distortion products. The output backoff necessary for TWTs is typically in the range 6dB to 10dB, although the available useful power output can be increased above that level by optimising the assignment of frequencies to carriers and by the use of TWT linearising networks. The second objective is to divide the available output carrier power between the carriers in accordance with their down-link transmission needs.

FDMA may be used for groups of carriers which have been modulated in any way, analogue or digital. Some of the carriers assigned frequencies in an FDMA system may themselves be multiple access systems, using time division multiple access (TDMA). Furthermore, if the C/N ratio in the output of the transponder is not too high, it may be feasible to overlay the FDMA signals with spread spectrum signals, forming, in effect, a code division multiple access (CDMA) system.

A time division multiple access (TDMA) system, operating alone in a transponder, allows the full power to the transponder to be used, that is, no backoff is required. This is because only one carrier is present in the transponder at any instant in time. Each Earth station in the system transmits its signals in turn, in bursts, in assigned time slots, typically using PSK modulation, a brief guard time being assigned between each pair of burst slots to ensure that the bursts do not overlap even if small timing errors arise. Figure 6.11 illustrates the frame structure of a high capacity TDMA system.

Signals which are to be transmitted over a TDMA system must be digital. Bits within a frame are stored at the transmitting Earth station, then assembled into a burst with the necessary preamble bits and transmitted at high speed at the appropriate time. At the receiver the reverse process puts the signal bits into store, then reads them out at the appropriate lower speed, frame by frame. The characteristics of TDMA systems vary over a wide range because the principle can be applied in many different circumstances, ranging from the transmission of low information rate monitoring or control signals with an aggregate bit rate of a few kbit/s, probably transmitted on a frequency assigned within a FDMA system, to the high capacity international telecommunications network TDMA systems operating at 120Mbit/s in the INTELSAT and EUTELSAT systems (INTELSAT, 1972; EUTELSAT, 1981; Hills and Evans, 1973).

On board switched TDMA has become feasible in multi-beam satellites like INTELSAT VI, using switch matrices which can operate within the TDMA frame to route one burst to down-link beam A and the next burst to another down-link beam, B.

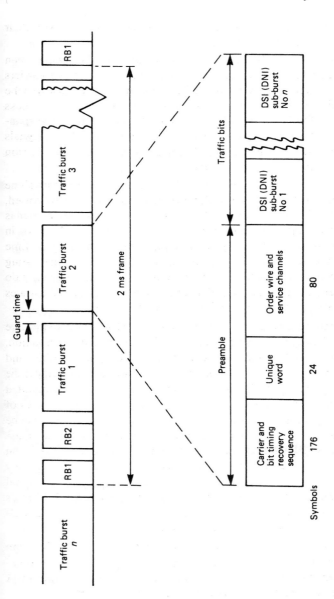

**Figure 6.11** Frame and burst format of the INTELSAT TDMA system. RB1 and RB2 are the reference bursts from reference stations 1 and 2 respectively. The drawing is not to scale

The functioning of TDMA systems which make efficient use of the time dimension demands precise timing and complex control of access. Such systems may be costly. Where the traffic flowing through the system is light, much simpler systems which use principles first explored within the Aloah system may provide adequate availability. In these, the transmission path is normally open and an Earth station with information to send verifies that no down-link burst from another Earth station is in progress; it then transmits its burst. However, several hundreds of milliseconds elapse before the start of a signal from an Earth station, sent via a geostationary satellite, can be received at another Earth station. Two Earth stations may therefore inadvertently transmit overlapping bursts, causing both messages to be mutilated. If this happens, they are both retransmitted automatically.

CDMA systems do not structure their use of transponders either in frequency or time. Earth stations transmit spread spectrum signals which can be identified, after re-transmission by the satellite, by the coding which the signal elements carry.

These various multiple access systems differ in the effectiveness with which they use the facilities provided by a transponder. Figure 6.12 provides a measure of the capacity of a transponder having a bandwidth of 36MHz, using various multiple access and modulation techniques, as a function of the C/N ratio at the Earth stations. Methods for calculating transponder performance are given in Hills and Evans, 1973, and in Bargellini, 1972.

# 6.5 Applications

Satellites can be used for virtually any kind of telecommunication system. They provide a very flexible transmission medium; once a satellite is available in orbit, many kinds of Earth station, for many kinds of use, can be put in place and set into operation very quickly. Satellite networks also offer high reliability. However, the costs of satellite systems fall quite differently when compared with the costs of terrestrial radio and cable systems where these are capable of providing equivalent facilities.

Communication satellite systems 209

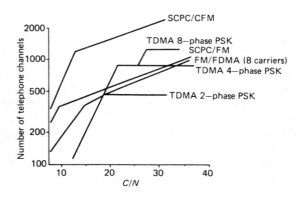

**Figure 6.12** Telephone channel capacity in 36 MHz channel

In some circumstances there are major differences between the demands which satellite and terrestrial radio systems make on another resource which is in short supply, namely the frequency spectrum. When these two factors, cost and spectrum availability, have been taken into account there are four main areas of application where satellite communication is already established or is becoming established and is likely to remain in significant use:

1. Trunk telecommunications. The global trunk network, linking fixed points at national centres and provincial centres all over the world, carrying massive amounts of telephone, data and video channels and all the other kinds of telecommunication facilities that can be conveyed by such channels, is routed mainly by terrestrial and submarine cable (increasingly optical fibre) and terrestrial radio relay systems. Such systems are very competitive in cost, relative to the satellite medium, for short distances. However, the distance between link terminals is sometimes large, and for these long links the satellite medium, the costs of which are little affected by distance, may be lower than those of terrestrial media.

2. Thin-route telecommunications. The economies of scale which make terrestrial transmission media so competitive for high capacity, short and medium distance links between pairs of fixed points do not necessarily apply in some other circumstances. Public telecommunications facilities may be provided economically by satellite over relatively short distances when the level of demand is low, especially where the intervening terrain is underdeveloped and rugged. Private telecommunications networks covering extensive geographical areas and requiring flexibility or non-standard facilities (such as high speed data and point to multipoint communication) are another area of application which is already important in some countries.

3. Communication with mobile stations. There is no terrestrial transmission medium for communication with ships and aircraft when far out of sight of land, which can be regarded nowadays as satisfactory. Satellite communication for ships is well established as the medium of choice. The use of satellites for communication with aircraft on ocean crossings has begun, and it is likely to be extended soon to air traffic control and other aeronautical communication services. Finally, while both cost and limitations of radio spectrum availability seem certain to ensure that most communication with land mobile stations, such as car and hand portable cellular radiotelephones, will be served by terrestrial radio systems, it is likely that satellite communication will extend the availability of such facilities into areas which would not be economically attractive for the establishment of terrestrial systems.

4. Broadcasting. Three transmission media, namely terrestrial radio, cable and satellite, are competing for domestic, standard definition, television broadcasting at present and each seems to be identifying markets where it can dominate. Studies are well advanced on two additional broadcasting services for the satellite medium, namely high definition television and high quality sound broadcasting suitable for reception in motor cars.

These four areas of application are considered further in the next section.

## 6.6 Satellite systems

### 6.6.1 Trunk telecommunications

The International Telecommunication Satellite Organisation (INTELSAT) is controlled in matters of policy by the many governments that are parties to the agreement that set it up and it is owned by the telecommunications undertakings that those governments have designated for the purpose. The INTELSAT system exhibits well the characteristics of a trunk telecommunications satellite system. The satellites, all geostationary, are located in three groups, for the Atlantic, Pacific and Indian Ocean regions. Some of the transponders are leased for domestic networks but the larger part of the system provides facilities for an international network linking Earth stations owned by national telecommunications operating organisations. It is characteristic of this global network that the total required capacity is very large; that large numbers of national Earth stations need access to it; that most of those Earth stations need to be able to communicate directly with many of the other Earth stations having access and that these country to country links vary greatly in scale, some requiring only one circuit but others requiring thousands of circuits (Hall and Moss, 1978; Alper and Pelton, 1984).

It would be ideal for INTELSAT users if a single satellite in each ocean region had enough fully interconnected capacity to carry all of the traffic that the region requires. Successive series of INTELSAT satellites have had larger capacities, satellite antenna beams have been kept wide to maximise connectivity and the required figure of merit (G/T) of Earth station antennas has been kept high to maximise per-satellite capacity. However, demand per region has outpaced capacity per satellite consistently since 1967. There are currently 13 operational satellites, of the INTELSAT V, VA and VI series, seven of them serving the Atlantic Ocean alone. In each ocean region, one satellite is designated as the Primary Satellite and every country is allocated some capacity in it, so that Earth stations with a relatively light total traffic flow can get access to all the routes of interest to them within their own ocean region without using more than one antenna. If the capacity required between two Earth stations is large, it will usually be carried by one of the other satellites serving the region, called a Major Path Satellite.

All current INTELSAT satellite series have transponders operating in both the 6GHz and 4GHz and the 14GHz and 11–12GHz bands. High gain spot beam dish antennas, steerable from the ground, are used for the transponders operating in the higher frequency band pair. For the lower band pair, some transponders use dual polar horn antennas covering the whole Earth visible from the satellite whilst other transponders use higher gain reconfigurable dual polar multi feed reflector antennas to cover smaller areas, the footprints being shaped to fit the geographical configuration of important service areas (see Figure 6.10). Both travelling wave tube and solid state power amplifiers are used with single carrier saturated output power ratings up to 20W, and transponder bandwidths ranging between 36MHz and 112MHz.

Earth station antennas having access to INTELSAT satellites for the global network are required to have high performance, to ensure that bandwidth and satellite power can be used efficiently. Until recently the standard receiving Figure of Merit (G/T) was 40.7dB/K at 4GHz and 39dB/K at 11GHz, requiring antennas with primary reflectors about 32 metres and 18 metres in diameter respectively. Many antennas now in service meet these requirements, but the minimum G/T requirements for new all purpose antenna has been reduced to 37dB/K (1GHz) and 35dB/K (11GHz). Much lower performance is acceptable for some limited applications (INTELSAT, 1977; Thompson and Buchsbaum, 1985).

The high gain antennas must use active satellite tracking techniques to avoid loss of performance due to satellite orbital perturbations; typically Earth station beam pointing errors are detected by observing the effect on the level of the telemetry beacon radiated by the satellite of small deliberate movements of the beam.

Both time division and frequency division multiple access methods are used in the INTELSAT global network for the transmission of narrow band channels which are designed primarily for telephony but which are, of course, also used for other narrow band telecommunications services. It is usual for the Earth station to combine the telephone channels it transmits to many destinations onto one or a few multiplexed carriers; conversely, a typical Earth station receives multiplex carriers from many other Earth stations, extracting from the multiplexes only those channels which are its concern.

The INTELSAT TDMA systems operate at an information rate of 120.832Mb/s, using 4-phase PSK modulation and coherent demodulation, with transponders of 72MHz bandwidth ((INTELSAT, 1972). Digital speech interpolation (DSI) is used to increase the capacity of the system for telephone traffic (Campanella, 1978). Satellite switched TDMA is available with INTELSAT VI satellites to enable up-beam or down-beam connectivity to be optimised.

However, most INTELSAT transponders are operated in the FDMA mode, with various modulation methods, analogue and digital. Multiplexed analogue systems use FM and FDM, with standardised basebands of various capacities, the structure of the baseband being related to the ITU-R (formerly CCIR) and ITU-T (formerly CCITT) standards for analogue radio relay system basebands. The deviation and power level of the carriers are optimised to give channels which attain their performance objectives with economical use of transponder power and modulator allotments of bandwidth. Most digital systems use TDM and 4-phase PSK, with channels 8-bit encoded, although single channel PSK systems are also used.

Another important function of the INTELSAT global network is to provide temporary, brief wide band links for the transmission of news, in the form of television pictures, from the scene of origin to all parts of the world, for re-broadcasting. A transponder is reserved for this purpose in each Primary Satellite and frequency modulation by an analogue video signal, which may contain a digitalised sound signal, time division multiplexed amongst the synchronising pulses, is usual.

Wide band low noise first stage amplifiers, typically parametric amplifiers, are used to complement the high receiving gain of the big Earth station antennas, and these amplifiers are usually thermo-electrically cooled (Peltier effect) at 4GHz. High power is often required from the output stages of Earth station transmitters, involving hundreds of watts output and sometimes several kilowatts, more especially when the station will be transmitting television signals to a global coverage transponder or where large numbers of telephone channels are to be transmitted on several carriers. These high power amplifier stages may consist of wide band travelling wave tubes, or groups of klystrons operating in parallel through frequency selective combining networks. In other respects the equipment of an Earth station accessing the INTELSAT global network is similar, in principle, to that of a major radio relay station.

The INTELSAT system is by far the biggest provider of satellite facilities for trunk telecommunications but the EUTELSAT system provides digital international links with Europe, closely resembling the TDMA system operated by INTELSAT and using very similar satellite and Earth station equipment (EUTELSAT, 1981). Similar configurations also arise in some of the North American domestic satellite networks. Recent satellite designs, like EUTELSAT II, tend to have transponders with higher output power, and a single carrier saturation rating of 50W is often found now.

For systems which are used in this way, the gain of transponders and such Earth station emission parameters as carrier power and (for FM) carrier deviation are typically determined so that the ITU-R and ITU-T channel noise and bit error ratio (BER) objectives for international telephone, data and other channels will be met (CCIR, 1980a; CCIR, 1980b; CCIR, 1980c). Foreseeable levels of interference from other satellite networks are allowed for, and a margin is left for interference from terrestrial radio systems operating in the same frequency bands. The more important of these criteria are as follows:

1. For analogue telephone channels, the total noise, including interference reckoned as noise, psophometrically weighted, at a point of zero relative level in the telephone network, shall not exceed 10000pW for more than 20% of the month. Nor shall this noise exceed 50000pW for more than 0.3% of any month.
2. Within (1), the margins allowed for interference from terrestrial radio systems are 1000pW for 20% of any month and 50000pW for 0.03% of any month (CCIR, 1980d).

3. For 64kbit/s digital channels which are to form part of an Integrated Services Digital Network (ISDN), the BER should not exceed one in $10^7$ for more than 10% of any month. For 2% and 0.03% of any month, the BER should not exceed one in $10^6$ and one in $10^3$ respectively.
4. For a video signal the objective is for a weighted signal to noise ratio of not worse than 53dB, for 99% of any month, measured in a baseband 0.01MHz to 5.0MHz.

### 6.6.2 Thin route telecommunications

INTELSAT and EUTELSAT transponders which are not required for trunk telecommunications may be leased to telecommunications operating organisations for national networks and for onward leasing to users for their private networks. A considerable number of other organisations have brought satellites into operation primarily to provide for national networks and private networks, and many more satellites are currently in the planning stage or under construction, their frequency assignments and orbital locations being co-ordinated with those of other satellites which are already in operation. These national satellites are broadly similar to those of INTELSAT and EUTELSAT. They use the same frequency bands, although there is a growing tendency to concentrate on the frequency bands which are allocated exclusively to the fixed satellite service (see Table 6.1) in order to avoid constraints arising from sharing spectrum with terrestrial radio systems. Their transponders are typically 36MHz in bandwidth and with a single carrier saturated power rating between 6 and 20W. They differ mainly in their use of antennas with higher gain and footprints which are limited, more or less, to national boundaries.

With few exceptions the Earth stations which use these satellite facilities are simple and relatively low in cost. Some have antennas of substantial size, up to 10m in diameter and requiring active satellite tracking but most antennas are less than 5m in diameter, making them less environmentally objectionable, and their relatively broad beams do not need to track actively a satellite which is in an accurately maintained geostationary orbit.

## 216  Satellite systems

These points of similarity apart, these Earth stations and the facilities that they provide show considerable variety. The following can perhaps be taken to cover the majority of cases:

1. In a number of large countries where high quality terrestrial communication facilities do not link all centres of population, satellite networks have been set up to provide trunk connections between provincial centres. Elsewhere, where the population is sparsely distributed over inaccessible terrain amongst mountains or across archipelagos, satellites may provide reliable communications to isolated communities. The trunk network configuration will usually resemble the INTELSAT global network, although on a smaller scale, using FDMA and with analogue telephone channels in frequency division multiplexed basebands frequency modulating the radio carriers, or digital channels in time division multiplexed phase shift modulating the carriers. Networks serving isolated communities are likely to use voice switched carriers for single telephone channels, analogue or digital.

2. Big corporations operating in well separated localities have found it worth while to lease satellite transponders for private telecommunications networks linking their establishments. In principle these networks resemble the national trunk network described in (1) above, but they may be more flexible in their configuration, being designed to cater for needs such as teleconferencing and high speed data transfer which are not well supplied at present by the public telecommunications networks.

3. The largest single use of satellite transponders in the fixed satellite service, at present, is for the distribution of television programme signals to terrestrial radio broadcasting stations and the head ends of cable broadcasting networks.

4. In recent years extensive use has developed, in particular in the USA, of very low cost Earth stations for data networks. These Earth stations use very small antennas, typically less than 2.5 metres in diameter; they are the so called Very Small

### 6.6.3 Satellite communication for mobile stations

An important advantage of satellite communication over all other media for long distance communication is its ability to operate with mobile terminals. Indeed, satellites provide the only feasible way of providing reliable communication with ships and aircraft which are far out of sight of land. The microwave frequencies which are used for communication between satellites and fixed Earth stations are, however, unsuitable for use with mobile stations. Lower frequencies provide a better trade off between the satellite carrier power required and the cost of mobile Earth station antennas. As shown in Table 6.2, frequency bands have been allocated for satellite mobile services around 1.6GHz and, in some parts of the world, at 800MHz. However, the 1.6GHz bands are narrow and the 800MHz bands are crowded by terrestrial services.

The first commercial satellite communication service for ships was provided in 1976 by the MARISAT satellites, owned by Comsat Corporation (Lipke et al., 1987). These satellites covered most significant sea areas, although with some gaps in arctic and antarctic waters. In 1982 the International Maritime Satellite Organisation, a global consortium similar to INTELSAT, took over and expanded this service. The INMARSAT system initially used the MARISAT satellites leased from COMSAT, some transponders which had been added for the purpose to three INTELSAT V satellites and two MARECS satellites on loan from the European Space Agency. All of these satellites operated their ship satellite links at 1.6GHz and their feeder links with Earth stations ashore at 6GHz and 4GHz. Their transponders had bandwidths of a few megahertz and the saturated output power of their transmitters was rated at a few tens of watts. All satellite antennas had footprints which covered the whole Earth visible from the satellite.

The standard INMARSAT ship Earth station antenna in the receive mode has a Figure of Merit (G/T) of −4dB/K, typically achieved by an antenna gain of 23dBi from a 1.2 metre dish and a receiver with a system noise temperature of 500K. Such an antenna has a −3dB beamwidth of about $10°$ and a simple beam stabilising and control system is sufficient to compensate for the movement of the ship and slow changes in the direction of the satellite as the ship changes its location.

Single channels of narrow bandwidth analogue telephony signals frequency modulate voice switched carriers, and the transponder power per channel requirement is further reduced by companding. A low speed TDMA system provides many channels of telex, for use with all equipped ships. A second TDMA system is used for signalling and order wire purposes to control both telephone and telex facilities.

A second generation of INMARSAT satellites is in the process of deployment, replacing the various satellites that brought the system into being (Berlin, 1986). These satellites operate in the same frequency bands as did their predecessors but the transponder bandwidth has been increased to 16MHz and the available down-link power is also considerably increased.

With the additional capacity that the second generation of satellites will bring to the INMARSAT system, it will be possible to provide for growth of maritime use but it also becomes possible to consider opening up new services. In 1987 the INMARSAT Convention was amended to permit the organisation to offer service to aircraft and a start has been made in offering connections to the public telephone network to passengers on some aircraft on trans-Atlantic flights. INMARSAT is also offering telex service to road vehicles.

Various other satellite systems are also being set up to provide communication services to mobile stations and in particular to road vehicles. Some of these will operate in the same frequency bands as INMARSAT. Others, and particularly some that are emerging in North America, will use the frequency allocations at 800MHz to 960 MHz.

### 6.6.4 Satellite broadcasting

Subject to various constraints to protect terrestrial radio services, and in particular terrestrial television broadcasting, which makes extensive use of the same band and has superior allocation status, para graph 693 of the ITU Radio Regulations permits satellite broadcasting in the band 620GHz to 790GHz. The USSR has been using EKRAN satellites to broadcast FM TV in this band for a number of years, providing signals that can be received over a wide area. However, there seems to be little prospect that satellite broadcasting will expand in this band.

The main frequency bands foreseen for satellite broadcasting are at 12GHz. Frequency assignment plans for broadcasting down-links in these bands, and the corresponding feeder links in other bands at 18GHz and 15GHz, were agreed at various administrative conferences of the ITU between 1977 and 1988. These plans define the ways in which the band is to be used in some considerable detail, but in essence:

1. With few exceptions, down-link footprints may not exceed national boundaries;
2. With allowances made for the climate of the country concerned, the power flux density at the ground at the edge of the coverage area should be $-103dB(W/m^2)$ for ITU Regions 1 and 3; $-107dB(W/m^2)$ for Region 2.
3. Every country has been assigned a share of the channels that the plan can provide. In Region 1 the shares are equal, every country being assigned 5 channels, except where the country is so large that provision must be made for extra channels to enable each time zone to have its own programmes.

These plans have been ratified by national governments, providing strong protection against cross frontier interference. Several countries had experimented with satellite broadcasting at 12 GHz before the plans were finalised (Siocos, 1978; Roscoe, 1980; Ishida et al., 1979). Others have launched satellites more recently to make use of the planned assignments, including TV-SAT, TDF-1 and Marco Polo for France, Germany and the United Kingdom, respectively. However, in general the take up of these planned assignments has been slow. The specified power flux density requires satellite transmitter output powers of several hundreds of watts for the larger countries, which causes capital costs to be high and it is arguable that recent

advances in the design of low cost satellite broadcasting receivers ha
made such a high power flux density unnecessary. Furthermore, the
limiting of coverage to national boundaries is no longer attractive in
for example, Western Europe. The general public is tending to get its
satellite broadcasting signals from other sources.

Many satellites of the fixed satellite service, in particular domestic
systems of the USA, are used to distribute substantial numbers of TV
programme channels to terrestrial radio stations, cable network head
ends, hotels and so on. However, members of the public also set up
domestic antennas to receive these transmissions for their own enjoy
ment. This process has spread, for example, to Europe, and accoun
has now been taken of it in the design of satellites. Thus, the ASTRA
satellites of SES and the EUTELSAT II series, both transmitting in
the 10.7GHz to 11.7GHz band, have relatively large beam footprint
for international coverage and are used, for example, to distribut
programmes to designated fixed Earth stations for onward distribu
tion to the public. However, the transponders of these satellites, being
equipped with 50W transmitters, provide a signal strength on th
ground which is not so very much less than the objective set for high
power satellites in the planned 12GHz bands; in consequence thes
satellites tend to be seen by the public as broadcasting satellites an
large numbers of homes treat them as such.

## 6.7 Future developments

For communication by satellite between fixed Earth stations th
distant future lies in a struggle, technical and economic, with terre
strial systems, cabled or by radio relay system. How fast will th
public telecommunications network become converted to a univer
sally available, omni-purpose, global, integrated, synchronised, digi
tal network? To what degree will satellites provide specia
communication facilities that the world is going to want and whic
are not available at an acceptable price from that integrated globa
network?

Who can tell? It seems likely, however, that in the nearer future th
use of satellites for main links in the public telecommunication
network will grow, as the world economy grows, but perhaps at

slower rate than hitherto and under strong pressure to minimise costs. On the other hand, there may well be considerable growth in the use of powerful satellites and small Earth stations for a variety of other purposes. Thin route public network systems will bring modern telecommunications facilities to sparsely populated and under developed areas. Private networks using VSATs may provide cost effective data transmission in some circumstances (Berger, 1994; Louvet and Chellingsworth, 1994; Pagni and Bosch, 1994).

Above all, use will be found for satellites for limited access multi-destination transmissions, 'narrow casting' as it is sometimes called, for conference television and for other situations where wide bandwidth is required for short periods. Indeed the indications are that the use of satellites for purpose of this kind may be limited, for the developed regions of the world, not by shortage of demand but by difficulties of supply, arising from increasing congestion in the geostationary satellite orbit.

These difficulties will, no doubt, lead to the use of the frequency bands allocated to the fixed satellite service around 20GHz and 30GHz (see Table 6.1) for applications where the additional cost of operating in these bands can be absorbed.

It is clear that there will be strong growth in demand for the use of satellites for communication with mobile stations, and in particular with road vehicles in sparsely populated regions. There will also be a considerable demand for communication by satellite with aircraft, to provide telephone services for passengers and probably also for air traffic control purposes.

Development of high definition television (HDTV) is in progress in many countries. Most of the systems that have been announced so far require much more information to be transmitted than is needed for standard definition TV. In consequence there are no ready means, terrestrial or satellite, of distributing such signals widely and there is interest in securing a new frequency allocation for satellite broadcasting of HDTV, probably around 20GHz. A second broadcasting development looks towards multi-channel sound broadcasting of 'CD-quality' that could be received in motor cars; for this purpose another new frequency allocation is being sought, but this would have to be much lower in the spectrum, probably below 3GHz.

## 6.8 Acknowledgements

The author wishes to thank the Commission of the European Communities for making this work possible and Mr D Withers for his assistance.

## 6.9 References

Alper, J. and Pelton, J. N. eds. (1984) The INTELSAT Global Satellite system, Progress in Astronautics and Aeronautics series Vol. 93 *American Institute of Aeronautics and Astronautics Inc.*

Bargellini, P.L. (1972) The INTELSAT IV communication system *COMSAT Tech. Rev.*, 2 **2** p. 437.

Berger, D.A. (1994) The time has arrived for VSATs in Europe *Telecommunications*, May.

Berlin, P. (1986) INMARSAT's second-generation satellites, *Proc IEE*, **133**, Part F. (4) p. 317.

Campanella, S. J. (1978) Digital speech interpolation techniques *1978 National Telecommunications Conf.*, Birmingham, Al USA, Conf. Record of the IEEE, 14.1/1.5 December.

CCIR (1980a) *Allowable noise power in the hypothetical reference circuit for frequency division multiplex telephony in the fixed-satellite service*, CCIR Rec. 353, ITU, Geneva.

CCIR (1980b) *Single value of the signal-to-noise ratio of all television systems*, CCIR Rec. 558, ITU Geneva.

CCIR (1980c) *Allowable bit error ratios at the output of the hypothetical digital reference path for systems of the fixed-satellite service using pulse code modulation for telephony*, CCIR Rec. 522 ITU Geneva.

CCIR (1980d) *Maximum allowable values of interference from line of-sight radio relay systems in a telephone channel of a system in the fixed satellite service employing frequency modulation, when the same frequency bands are shared by both systems*, CCIR Rec. 356, ITU Geneva.

Chitre, N.K.M. and Fuenzalida, J.C. (1972) Baseband distortion caused by intermodulation in multicarrier FM systems, *COMSAT Tech. Rev,* 2, **1** p. 147.

Clarke, A.C. (1945) Extra-terrestrial relays, *Wireless World*, p. 305.

Colcy, J.N., et al. (1995) Euteltracs: the European mobile satellite service, *Electronics & Communication Engineering Journal*, April.

Communicate (1995) The race for world domination heats up, *Communicate*, May.

Comparettto G.M. and Holkower, N.D. (1995) More than just a mirage on the horizon, *Mobile Europe*, January.

Dickieson, A.C. (1963) The TELSTAR experiment, *Bell Syst. Tech. J*, **42** p. 739, (and associated papers).

EUTELSAT (1981) TDMA/DSI system specification, *EUTELSAT Document ESC/C-11-17 Rev 1*, September.

Evans, B.G. et al. (1986) Baseband switches and transmultiplexers for use in an on-board processing mobile/business satellite system, *Proc. IEE*, Part F, p. 356.

Free, B.A. (1980) North-south station keeping with electric propulsion using on-board battery power, *JANNUF Propulsion Meeting*, **5** p. 217.

Fuenzalida, J.C., Rivalan, P. and Weiss H.J. (1977) Summary of the INTELSAT V communications performance specifications, *COMSAT Tech. Rev.* **7** No.1 p. 311.

Hall, G.C. and Moss P.R. (1978) A review of the development of the INTELSAT system, *Post Office Elect. Eng. J,* **71** p. 155.

Hayn, D., Braitinger, M. and Schmucker, R.H. (1978) Performance prediction of power augmented electrothermal hydrazine thrusters, *Technische Universitaet Lehrstuhl fur Raumfahrttechnik*, Munich, W. Germany.

Hills, M.T. and Evans B.G. (1973) *Telecommunications System Design*, Vol.1 Transmission Systems, pp. 176-198, George Allanand Unwin.

INTELSAT (1972) TDMA/DSI system specification (TDMA/DSI Traffic terminals), *INTELSAT Document BG 42/65 (Rev1)* June 1981.

INTELSAT (1977) *Standard A performance characteristics of Earth stations in the INTELSAT IV, IVA and V systems having a G/T of 40.7dB/K (6.4GHz frequency bands)*, INTELSAT document BG 28/72, 75, August.

Ishida, et al. (1979) Present situation of Japanese satellite broadcasting for experimental purposes, *IEEE Trans,* **BC-25**, (5) p. 105.

ITU (1990a), Article 8, *Radio Regulations,* International Telecommunication Union, Geneva.

ITU (1990b) Appendix 30B, *Radio Regulations*.

ITU (1990c) Para. 837, *Radio Regulations*.

ITU (1990d) Para. 700, *Radio Regulations*.

ITU (1990e) Para. 701, *Radio Regulations*.

ITU (1990f) Para. 726B, *Radio Regulations*.

ITU (1990g) Appendix 30, *Radio Regulations*.

ITU (1990h) Appendix 30A, *Radio Regulations*.

ITU (1990i) Para. 2613, *Radio Regulations*.

ITU (1990j) Articles 11 and 13 and Appendix 29, *Radio Regulations*.

ITU (1990k) Paras. 2552 – 2585, *Radio Regulations*.

ITU (1990l) Articles 11, 12 and 13 and Appendix 28, *Radio Regulations*.

ITU (1990m) Paras. 2615 – 2617 and 2619, *Radio Regulations*.

ITU (1990n) Paras. 2618 and 2620 – 2623, *Radio Regulations*.

ITU (1990o) Paras. 2624 – 2627, *Radio Regulations*.

ITU (1990p) Paras, 2628 – 2630, *Radio Regulations*.

Jilg, E.T. (1972) The INTELSAT IV Spacecraft, *COMSAT Tech. Rev.*, **2** No.2, p. 271.

Libbenga, J. (1995) Satellite spangled space, *Mobile Europe*, May.

Lipke, D.W. et al., (1987) MARISAT — a maritime satellite communications system, *COMSAT Tech. Rev.* **7**, (2) p. 36.

Louvet, B. and Chellingsworth, S. (1994) Satellite integration into broadband networks, *Electrical Communication*, 3rd Quarter.

Mann, K. (1995) Into the mainstream, *Communications News*, January.

Nabet, D. (1995) The Global broadcast industry: the benefits for telecoms, *Telecommunications*, January.

NASA (1968) *Relay program final report*, NASA special publication SP-16.

Pagni, L. and Guillermo, B. (1994) VSATs: integrated communications and services, *Telecommunications*, November.

Roscoe, O.S. (1980) Direct broadcasting satellites — the Canadian experience, *Satell. Commun*, (USA) **4** (8) p. 22.

Sachdev, D.K. (1990) Historical overview of the INTELSAT system, *J. Br. Interplanet. Soc.* **43**, p. 331, August.

Schneiderman, R. (1994) Direct-broadcast satellites make home deliveries, *Microwaves & RF*, July.

Siocos, C.A. (1978) Broadcasting satellite reception experiment in Canada using high power satellite HERMES, IBC-78, *IEE Conf. Publ.*, (UK) **166**, p. 197.

Thompson, P.T. and Buchsbaum. L.M. (1985) INTELSAT Earth-station standards — A new look to an old theme, *International Journal of Satellite Communications*, **3**, p. 259.

Watt, N. (1986) Multibeam SS-TDMA design considerations relating to the Olympus Specialised Services Payload, *Proc. IEE 133*, Part F, **4** p. 317.

Westcott, R.J. (1972) Investigation of multiple FM/FDM carriers through a satellite TWT operating near to saturation, *Proc. IEEE*, 114, **6** p. 726.

White, I. (1995) Space Odyssey, *Mobile News*, 20 February.

# 7. Acronyms

Every discipline has its own 'language' and this is especially true of telecommunications, where acronyms abound. In this guide to acronyms, where the letters within an acronym can have slightly different interpretations, these are given within the same entry. If the acronym stands for completely different terms then these are listed separately.

| | |
|---|---|
| AAN | All Area Networking. (Networking covering local and wide areas. Also used to imply combined use of LAN and WAN.) |
| ACK | Acknowledgement. (Control code sent from a receiver to a transmitter to acknowledge the receipt of a transmission.) |
| ACS | Accredited Standards Committee. (ANSI.) |
| ACTE | Approvals Committee for Terminal Equipment. (Part of European Community.) |
| ACTIUS | Association of CTI Users and Suppliers. |
| ADC | Analogue to Digital Conversion. |
| ADC | Access Deficit Contributions. |
| ADM | Adaptive Delta Modulation. (Digital signal modulation technique.) |
| ADM | Add-Drop Multiplexer. (Term sometimes used to describe a drop and insert multiplexer.) |
| ADP | Automatic Data Processing. |
| ADPCM | Adaptive Differential Pulse Code Modulation. (ITU-T standard for the conversion and transmission of analogue signals at 32kbit/s.) |
| AE | Anomaly Events. (E.g. frame errors, parity errors, etc. ITU-T M.550 for digital circuit testing.) |
| AFC | Automatic Frequency Control. |

| | |
|---|---|
| AFIPS | American Federation of Information Processing Societies. |
| AFNOR | Association Francaise de Normalisation. (Standardisation body in France.) |
| AFRTS | American Forces Radio and Television Services. |
| AGC | Automatic Gain Control. |
| AIIA | Australian Information Industry Association. |
| AIN | Advanced Intelligent Network. (Bellcore released specification for provision of wide range of telecommunication capabilities and services.) |
| AM | Amplitude Modulation. (Analogue signal transmission encoding technique.) |
| AMA | Automatic Message Accounting. (Ability within an office to automatically record call information for accounting purposes. See also CAMA.) |
| AMI | Alternate Mark Inversion. (Line code system.) |
| AMPS | Advanced Mobile Phone System. (Or American Mobile Phone Standard. Analogue cellular radio standard.) |
| ANBFM | Adaptive Narrow Band Frequency Modulation. |
| ANBS | American National Bureau of Standards. |
| ANI | Automatic Number Identification. (Feature for automatically determining the identity of the caller.) |
| ANSI | American National Standards Institute. |
| AOS | Alternate Operator Services. (US companies providing services in competition with existing suppliers such as AT&T and the RBOCs.) |
| AOTC | Australian and Overseas Telecommunications Corporation. (Australian PTT.) |
| APC | Aeronautical Public Correspondence. (ITU-T term for airborne communication systems.) |
| APC | Adaptive Predictive Coding. |
| API | Application Programming Interface. |
| APK | Amplitude Phase Keying. (A digital modulation technique in which the amplitude and phase of the carrier are varied.) |
| APNSS | Analogue Private Network Signalling System. |

| | |
|---|---|
| ARPA | Advanced Research Project Agency. (Agency operating within the US Department of Defence.) |
| ARQ | Automatic Request for repetition. (A feature in transmission systems in which the receiver automatically asks the sender to retransmit a block of information, usually because there is an error in the earlier transmission.) |
| ASCII | American Standard Code for Information Interchange. (Popular character code used for data communications and processing. Consists of seven bits, or eight bits with a parity bit added.) |
| ASIC | Application Specific Integrated Circuit. (Integrated circuit components which can be readily customised for a given application.) |
| ASK | Amplitude Shift Keying. (Digital modulation technique.) |
| ATDM | Asynchronous Time Division Multiplexing. |
| ATM | Asynchronous Transfer Mode. (ITU-T protocol for the transmission of voice, data and video.) |
| ATUG | Australian Telecommunications Users' Group. |
| ATV | Advanced Television. |
| AUSSAT | Australian national Satellite system operating company. |
| AUSTEL | Australian Telecommunications authority. |
| AVDM | Analogue Variable Delta Modulation. |
| AVL | Automatic Vehicle Location. |
| AWG | American Wire Gauge. |
| | |
| B-ISDN | Broadband Integrated Services Digital Network. |
| BABT | British Approvals Board for Telecommunications. |
| BBER | Background Block Error Ratio. |
| BCC | Block Check Character. (A control character which is added to a block of transmitted data, used in checking for errors.) |
| BCD | Binary Coded Decimal. (An older character code set, in which numbers are represented by a four bit sequence.) |

| | |
|---|---|
| BCH | Bose Chaudhure Hocquengherm. (Coding technique.) |
| BDT | Bureau for Development of Telecommunications. (Part of the ITU.) |
| BDF | Building Distribution Frame. |
| BELLCORE | Bell Communications Research. (Research organisation, incorporating parts of the former Bell Laboratories, established after the divestiture of AT&T. Funded by the BOCs and RBOCs to formulate telecommunication standards.) |
| BER | Bit Error Ratio. (Also called Bit Error Rate. It is a measure of transmission quality. It is the number of bits received in error during a transmission, divided by the total number of bits transmitted in a specific interval.) |
| BERT | Bit Error Ratio Tester. (Equipment used for digital transmission testing.) |
| BETRS | Basic Exchange Telecommunications Radio Service. (FCC) |
| BEXR | Basic Exchange Radio. |
| BIP | Bit Interleaved Parity. (A simple method of parity checking.) |
| BIST | Built In Self Test. |
| BISYNC | Binary Synchronous communications. (Older protocol used for character oriented transmission on half-duplex links.) |
| BMPT | Bundesministerium fur Post und Telekommunikation. (German telecommunication regulator.) |
| BnZS | Bipolar with n-Zero Substitution. (A channel code. Examples are B3ZS which has three-zero substitution; B6ZS with six-zero substitution, etc.) |
| BOC | Bell Operating Company. (Twenty-two BOCs were formed after the divestiture of AT&T, acting as local telephone companies in the US. They are now organised into seven Regional Bell Operating Companies or RBOCs.) |
| BPSK | Binary Phase Shift Keying. |

| | |
|---|---|
| BPV | Bipolar Violation. (Impairment of digital transmission system, using bipolar coding, where two pulses occur consecutively with the same polarity.) |
| BRA | Basic Rate Access. (ISDN, 2B+D code.) |
| BRAP | Broadcast Recognition with Alternating Priorities. (Multiple access technique.) |
| BRITE | Basic Research in Industrial Technology for Europe. (EC programme.) |
| BRZ | Bipolar Return to Zero. (A channel coding technique, used for digital transmission.) |
| BS | Base Station. (Used in mobile radio systems.) |
| BSC | Base Station Controller. (Controllers for Base Transceiver Stations, or BTS, within mobile radio systems.) |
| BSGL | Branch System General Licence. (Telecommunications licence in the UK.) |
| BSI | British Standards Institute. |
| BSM | Broadband Switched Mass-market. |
| BSS | Broadcast Satellite Service. |
| BT | Formerly British Telecom. (UK.) |
| BTS | Base Transceiver Station. (Used in mobile radio based systems to provide the air interface to the customer.) |
| BTV | Business TeleVision. |
| BUNI | Broadband User Network Interface. |
| | |
| CAC | Comite d'Action Commerciale. (Commercial Action Committee. Part of CEPT.) |
| CAI | Common Air Interface. (Radio communication standard supported by ETSI.) |
| CAMA | Centralised Automatic Message Acconting. (A centralised version of AMA, used in larger offices serving several smaller ones which may be too small to justify AMA on their own.) |
| CANTO | Caribbean Association of National Telecommunication Organisations. (Association of state owned or |

private telecommunications carriers providing domestic and international services in the Caribbean.)

| | |
|---|---|
| CAP | Competitive Access Provider. (USA.) |
| CAPCS | Cellular Auxiliary Personal Communications Service. (TIA/EIA/IS-94 standard.) |
| CARS | Community Antenna Radio Service. |
| CAS | Channel Associated Signalling. (ITU-T signalling method.) |
| CATA | Community Antenna Television Association. |
| CATV | Community Antenna Television. (Also referred to as Cable Television.) |
| CBC | Canadian Broadcasting Corporation. |
| CBCS | Cordless Business Communication System. |
| CBDS | Connectionless Broadband Data Service. (ETSI's version of SMDS.) |
| CBEMA | Computer Business Equipment Manufacturers' Association. (USA.) |
| CBX | Computer controlled PBX. |
| CCA | Cable Communications Association. (Association of cable television and telephony providers. Formerly known as the Cable Television Association.) |
| CCC | Clear Channel Capacity. |
| CCH | Comite de Co-ordination de Harmonisation. (CEPT) |
| CCI | Co-channel Interference. (Interference between two subscribers, using the same channel but in different cells, of a cellular mobile radio system.) |
| CCIR | Comite Consultatif Internationale des Radiocommunications. (International Radio Consultative Committee. Former standards making body within the ITU and now part of its new Radiocommunication Sector.) |
| CCIS | Common Channel Interoffice Signalling. (North American signalling system which uses a separate signalling network between switches.) |
| CCITT | Comite Consultatif Internationale de Telephonique et Telegraphique. (Consulative Committee for Inter- |

| | national Telephone and Telegraphy. Standards making body within the ITU, now forming part of the new Standardisation Sector.) |
|---|---|
| CCR | Commitment, Concurrency and Recovery. |
| CCS | Common Channel Signalling. (ITU-T standard signalling system. Also called CCSS.) |
| CCSS | Common Channel Signalling System. (ITU-T standard signalling system. Also called CCS or No 7 signalling.) |
| CCTA | Central Communications and Telecommunications Agency. (UK Government equipment procurement agency.) |
| CCTS | Comite de Co-ordination pour les Telecommunications par Satellites. (Co-ordination committee for satellite telecommunications. CEPT.) |
| CCV | Co-ordination Committee for Vocabulary. (ITU-T/IEC.) |
| CDM | Code Delta Modulation. (Or Continuous Delta Modulation.) |
| CDMA | Code Division Multiple Access. |
| CDO | Community Dial Office. (Usually refers to an unattended switching centre which serves a small community and is controlled from a larger central office.) |
| CDPD | Cellular Digital Packet Data. (Data transmission technique using cellular voice networks.) |
| CEC | Commission of the European Communities. |
| CEE | Central and Eastern Europe. |
| CEN | Comite Europeen de Normalisation. (Committee for European Standardisation.) |
| CENELEC | Comite Europeen de Normalisation Electrotechnique. (Committee for European Electrotechnical Standardisation.) |
| CEPT | Conference des administrations Europeenes des Postes et Telecommunications. (Conference of European Posts and Telecommunications administrations. Body representing European PTTs.) |

## Acronyms 233

| | |
|---|---|
| CFM | Companded Frequency Modulation. |
| CFM | Composite Fade Margin. (Used in microwave system design.) |
| CFSK | Coherent Frequency Shift Keying. |
| CICC | Contactless Integrated Circuit Card. (ISO standard for a smart card in which information is stored and retrieved without use of conductive contacts.) |
| CIE | Commission Internationale de l'Eclairage. |
| CIS | Commonwealth of Independent States. (Alliance of states following the collapse of the USSR.) |
| CISPR | Comite International Special des Perturbations Radioelectriques. (IEC International Special Committee on Radio Interference.) |
| CIT | Computer Integrated Telephony. |
| CLAN | Cableless Local Area Network. (Radio based LAN.) |
| CLASS | Custom Local Area Signalling Services. |
| CLI | Command Line Interface. (Usually refers to an interface which allows remote asynchronous terminal access into a network management system. Also referred to as Command Line Interpreter.) |
| CLI | Calling Line Identity. (Facility for determining identity of the caller. See also CLID.) |
| CLID | Calling Line Identification. (Telephone facility which allows the called party to determine the identity of the caller.) |
| CLTA | Comite de liaison pour les telecommunications transatlantiques. (Liaison committee for transatlantic telecommunications. CEPT.) |
| CLTS | Connection-Less Transport Service. |
| CMI | Code Mark Inversion. (Line coding technique.) |
| CMS | Call Management System. |
| CMT | Character Mode Terminal. (E.g. VT100, which does not provide graphical capability.) |
| CNLP | Connection-Less Network Protocol. (Same as CLIP.) |
| CO | Central Office. (Usually refers a central switching or control centre belonging to a PTT.) |

## 234 Acronyms

| | |
|---|---|
| CODEC | COder-DECoder. |
| COMSAT | Communication Satellite Corporation. |
| | Customer Premise Equipment. |
| CPFSK | Continuous Phase Frequency Shift Keying. |
| CPODA | Contention Priority-Oriented Demand Assignment protocol. (Multiple access technique with contention for reservations. See PODA and FPODA.) |
| CPSK | Coherent Phase Shift Keying. |
| CPU | Central Processing Unit. (Usually part of a computer.) |
| CRA | Call Routing Apparatus. |
| CRC | Cyclic Redundancy Check. (Bit oriented protocol used for checking for errors in transmitted data.) |
| CRTC | Canadian Radio, television and Telecommunications Commision. (Canadian government telecommunication regulatory body.) |
| CS | Central Station. |
| CSBS | Customer Support and Billing System. |
| CSDN | Circuit Switched Data Network. |
| CSMA | Carrier Sense Multiple Access. (LAN multiple access technique.) |
| CSMA/CD | Carrier Sense Multiple Access with Collision Detection. (LAN access technique, with improved throughput, under heavy load conditions, compared to pure CSMA.) |
| CSTA | Computer Supported Telephony Applications. (Or Computer Supported Telecommunication Applications. For example, telemarketing applications.) |
| CT | Cordless Telephony. (CT1 is first generation; CT2 is second generation, and CT3 is third generation.) |
| CTA | Cable Television Association. (UK trade association of cable television suppliers, now renamed Cable Communications Association.) |
| CTCSS | Continuous Tone Controlled Squelch System. (Method for calling in paging systems.) |
| CTD | Centre for Telecommunications Development. (Part of ITU.) |

| | |
|---|---|
| CTIA | Cellular Telecommunications Industry Association. (USA trade association.) |
| CTR | Common Technical Regulation. (European Community mandatory standard.) |
| CTS | Conformance Test Services. (Part of European Commission initiative.) |
| CVSD | Continuous Variable Slope Delta modulation. (Proprietary method used for speech compression. Also called CVSDM.) |
| CW | Continuous Wave. |
| | |
| DAB | Digital Audio Broadcasting. |
| DAMA | Demand Assigned Multiple Access. |
| DAMPS | Digital Advanced Mobile Phone System. (Digital version of AMPS, which has the same basic architecture and signalling protocol.) |
| DAR | Dynamic Alternative Routing. (Traffic routing scheme proposed by BT in the UK.) |
| DARPA | Defence Advanced Research Projects Agency. (USA Government agency.) |
| DASS | Digital Access Signalling System. (Signalling system introduced in the UK prior to ITU-T standards I.440 and I.450.) |
| DASS | Demand Assignment Signalling and Switching unit. |
| DATTS | Data Acquisition Telecommand and Tracking Station. |
| DATV | Digitally Assisted Television. |
| DBS | Direct Broadcasting by Satellite. (Or Direct Broadcast Satellite. High power satellite suitable for directly beaming television programmes onto small receive-only satellite dishes.) |
| DBT | Deutsche Bunderspost Telekom. (Also written DBP-T. German PTT.) |
| DCA | Dynamic Channel Allocation. (System in which the operating frequency is selected by the equipment at time of use, rather than by a planned assignment.) |
| DCC | Data Communication Channel. |

## 236 Acronyms

| | |
|---|---|
| DCDM | Digitally Coded Delta Modulation. (Modulation technique where step size is controlled by sequence produced by the sampling and quantisation.) |
| DCE | Data Circuit termination Equipment. (Exchange end of a network, connecting to a DTE. Usually used in packet switched networks.) |
| DCF | Data Communications Function. |
| DCN | Data Communications Network. |
| DCR | Dynamically Controlled Routing. (Routing method proposed by Bell Northern Research, Canada.) |
| DCS | Digital Cellular System. (For example DCS 1800 operating at 1800MHz.) |
| DCS | Digital Crossconnect System. (See also DCX and DXC.) |
| DCS | Dynamic Channel Selection. (For example as used on call set-up in the CEPT 900MHz analogue cordless telephony standard.) |
| DCT | Discrete Cosine Transform. (Technique used in transform picture coding.) |
| DCX | Digital Crossconnect. (See also DXC and DSC.) |
| DDC | Data Country Code. (Part of an international telephone number.) |
| DDD | Direct Distance Dialling. (Generally refers to conventional dial-up long distance calls placed over a telephone network without operator assistance.) |
| DDN | Digital Data Network. |
| DDN | Defence Data Network. (US military network, derived from the ARPANET.) |
| DDP | Distributed Data Processing. |
| DDS | Digital Data Service. (North American data service.) |
| DE | Defect Events. (E.g. loss of signal, loss of frame synchronisation, etc. ITU-T M.550 for digital circuit testing.) |
| DECT | Digital European Cordless Telephony. (Or Digital European Cordless Telecommunications. ETSI standard, intended to be a replacement for CT2.) |

| | |
|---|---|
| DEDM | Dolby Enhanced Delta Modulation. |
| DELTA | Developing European Learning through Technological Advance. (European Community information technology and telecommunication programme associated with training.) |
| DEPSK | Differentially Encoded Phase Shift Keying. |
| DES | Data Encryption Standard. (Public standard encryption system from the American National Bureau of Standards.) |
| DFM | Dispersive Fade Margin. (Used in microwave system design.) |
| DFT | Discrete Fourier Transform. |
| DG | Directorates-General. (Departments of the European Commission, usually referred to by numerals, e.g. DGIV.) |
| DGPT | Direction General des Postes et Telecommunications. (France's national regulatory authority.) |
| DIN | Deutsches Institute fur Normung. (Standardisation body in Germany.) |
| DIS | Draft International Standard. |
| DM | Delta Modulation. (Digital signal modulation technique.) |
| DNIC | Data Network Identification Code. (Part of an international telephone number.) |
| DOD | Department Of Defence. (US agency.) |
| DOTAC | Department of Transport And Communications. (Australian.) |
| DPCM | Differential Pulse Code Modulation. |
| DPNSS | Digital Private Network Signalling System. (Inter-PABX signalling system used in the UK.) |
| DQDB | Distributed Queue Double Bus. (IEEE standard 802.6 for Metropolitan Area Networks.) |
| DQPSK | Differential Quaternary Phase Shift Keying. |
| DRG | Direction a la Reglementation Generale. (Directorate for General Regulation, in France.) |
| DRIVE | Dedicated Road Infrastructure for Vehicle safety in Europe. (European Community programme in the |

| | |
|---|---|
| | area of information technology and telecommunications, associated with road transport.) |
| DS-0 | Digital Signal level 0. (Part of the US transmission hierarchy, transmitting at 64kbit/s. DS-1 transmits at 1.544Mbit/s, DS-2 at 6.312Mbit/s, etc.) |
| DSB | Double Sideband. |
| DSBEC | Double Sideband Emitted Carrier. |
| DSBSC | Double Sideband Suppressed Carrier modulation. (A method for amplitude modulation of a signal.) |
| DSC | District Switching Centre. (Part of the switching hierarchy in BT's network.) |
| DSE | Data Switching Exchange. (Part of packet switched network.) |
| DSI | Digital Speech Interpolation. (Method used in digital speech transmission where the channel is activated only when speech is present.) |
| DSM | Delta Sigma Modulation. (Digital signal modulation technique.) |
| DSP | Digital Signal Processing. |
| DSRR | Digital Short-Range Radio. |
| DSS | Digital Subscriber Signalling. (CCIT term for the N-ISDN access protocol.) |
| DSSS | Direct Sequence Spread Spectrum. |
| DSU | Data Service Unit. (Customer premise interface to a digital line provided by a PTT.) |
| DSX-1 | Digital Signal Crossconnect. (Crossconnect used for DS-1 signals.) |
| DTE | Data Terminal Equipment. (User end of network which connects to a DCE. Usually used in packet switched networks.) |
| DTH | Direct to Home. (Usually refers to satellite TV.) |
| DTI | Department of Trade and Industry. |
| DTMF | Dual Tone Multi-Frequency. (Telephone signalling system used with push button telephones.) |
| DTS | Digital Termination System. (Local radio loop provided by carriers in the US.) |
| DXC | Digital Crossconnect. (See also DCX and DCS.) |

| | |
|---|---|
| EAU | European Unit of Account. (Earlier monitory unit of the EEC which was created in 1974 and replaced by the ECU in 1981.) |
| EB | Errored Block. (Measurement of transmission errors.) |
| EBCDIC | Extended Binary Coded Decimal Interchange Code. (Eight bit character code set.) |
| EBRD | European Bank for Reconstruction and Development. |
| EBU | European Broadcasting Union. |
| EC | European Commission. |
| ECC | Error Control Coding. (Coding used to reduce errors in transmission.) |
| ECITC | European Committee for IT Certification. |
| ECJ | European Court of Justice. |
| ECMA | European Computer Manufacturers Association. |
| ECREEA | European Conference of Radio and Electronic Equipment Association. |
| ECS | European Communication Satellite. |
| ECSA | Exchange Carriers Standards Association. (USA.) |
| ECTEL | European Conference of Telecommunications managers. |
| ECTF | Enterprise Computer Telephony Forum. |
| ECTRA | European Committee for Telecommunications Regulatory Affairs. (Part of CEPT.) |
| ECTUA | European Council of Telecommunications Users' Associations. |
| ECU | European Currency Unit. (Monetary unit of the EEC, created in 1981.) |
| EDI | Electronic Data Interchange. (Protocol for interchanging data between computer based systems.) |
| EDIFACT | EDI For Administration Commerce and Transport. (International rules for trading documents, e.g. purchase orders, payment orders, etc.) |
| EEA | Electrical Engineering Association. |
| EEA | Electronic and business Equipment Association. |
| EEC | European Economic Community. |

| | |
|---|---|
| EEMA | European Electronic Mail Association. |
| EFS | Error Free Seconds. (In transmitted data it determines the proportion of one second intervals, over a given period, when the data is error free.) |
| EFT | Electronic Funds Transfer. |
| EFTA | European Free Trade Association. |
| EFTPOS | Electronic Funds Transfer at the Point Of Sale. |
| EHF | Extremely High Frequency. (Usually used to describe the portion of the electromagnetic spectrum in the range 30GHz to 300GHz.) |
| EIA | Electronic Industries Association. (Trade association in USA.) |
| EIRP | Equivalent Isotropically Radiated Power. (Or Effective Isotropically Radiated Power. Of an antenna.) |
| ELT | Emergency Locator Transmitter. |
| EMA | Electronic Mail Association. |
| E-MAIL | Electronic Mail. |
| EMC | Electromagnetic Compatibility. |
| EMI | Electromagnetic Interference. |
| EMP | ElectroMagnetic Pulse. (Released by a nuclear explosion.) |
| EMS | European Monetary System. |
| EMUG | European MAP Users' Group. |
| EN | Equipment Number. (Code given to a line circuit, primarily in switches, to indicate its location on equipment racks.) |
| EN | Europaische Norm. (Norm Europeenne or European standard.) |
| ENV | European pre-standard. |
| EOA | End Of Address. (Header code used in a transmitted frame.) |
| EOB | End Of Block. (Character used at end of a transmitted frame. Also referred to as End of Transmitted Block or ETB.) |
| EOC | Embedded Operations Channel. (Bits carried in a transmission frame which contain auxiliary infor- |

| | |
|---|---|
| | mation such as for maintenance and supervisory. This is also called a Facilities Data Link, FDL.) |
| EOT | End Of Transmission. (Control code used in transmission to signal the receiver that all the information has been sent.) |
| EOTC | European Organisation for Testing and Certification. |
| EOTT | End Office Toll Trunking. (US term for trunks which are located between end offices situated in different toll areas.) |
| EOW | Engineering Order Wire. (A channel for voice or data communication between two stations on a transmission line.) |
| EP | European Parliment. |
| EPHOS | European Procurement Handbook for Open Systems. (Equivalent to GOSIP.) |
| EPIRB | Emergency Position Indicating Radio Beacon. |
| ERC | European Radiocommunications Committee. (Part of CEPT.) |
| ERDF | European Regional Development Fund. |
| ERL | Echo Return Loss. |
| ERM | Exchange Rate Mechanism. (Used within the European Community.) |
| ERMES | European Radio Messaging Service. (Pan-European standard for paging.) |
| ERO | European Radiocommunications Office. |
| ERP | Equipment Radiated Power. (Also referred to as Effective Radiated Power of an antenna.) |
| ES | Errored Second. (Measurement of transmission errors.) |
| ESA | European Space Agency. |
| ESB | Emergency Service Bureau. (A centralised location to which all emergency calls (e.g. police, ambulance, fire brigade) are routed. |
| ESF | Extended Superframe. (North American 24 frame digital transmission format.) |

| | |
|---|---|
| ESN | Electronic Serial Number. (Usually refers to the personal identity number coded into mobile radio handsets.) |
| ESO | European Standardisation Organisation. |
| ESPA | European Selective Paging Association. |
| ESPRIT | European Strategic Programme for Research and Development in Information Technologies. |
| ESR | Errored Second Ratio. (Measurement of transmission errors.) |
| ESS | Electronic Switching System. (A generic term used to describe stored programme control exchange switching systems.) |
| ETACS | Extended Total Access Communications System. (Extension of TACS with additional channel allocation below the existing TACS channels.) |
| ETB | End of Transmission Block. (A control character which denotes the end of a block of Bisync transmitted data.) |
| ETCO | European Telecommunications Consultancy Organisation. |
| ETE | Exchange Terminating Equipment. |
| ETIS | European Telecommunications Information Services. (Part of CEPT.) |
| ETNO | European Telecommunications Network Operators. (Association of European public operators.) |
| ETS | Electronic Tandem Switch. |
| ETS | European Telecommunication Standard. (Norme Europeenne de Telecommunications. Standard produced by ETSI.) |
| ETSI | European Telecommunications Standards Institute. |
| ETX | End of Text. (A control character used to denote the end of transmitted text, which was started by a STX character.) |
| EUCATEL | European Conference of Associations of Telecommunication industries.) |
| EURESCOM | European Institute for Research and Strategic Studies in Communications. (Part of CEPT.) |

| | |
|---|---|
| EUTELSAT | European Telecommunications Satellite organisation. |
| EVUA | European VPN Users Association. |
| FAS | Frame Alignment Signal. (Used in the alignment of digital transmission frames.) |
| FCC | Federal Communications Commission. (US authority, appointed by the President to regulate all interstate and international telecommunications.) |
| FCS | Frame Check Sequence. (Field added to a transmitted frame to check for errors.) |
| FDL | Facilities Data Link. (See EOC.) |
| FDM | Frequency Division Multiplexing. (Signal multiplexing technique.) |
| FDMA | Frequency Division Multiple Access. (Multiple access technique based on FDM.) |
| FDX | Full Duplex. (Transmission system in which the two stations connected by a link can transmit and receive simultaneously.) |
| FEC | Feedforward Error Correction. (Also called Forward Error Correction. Technique for correcting errors due to transmission.) |
| FEXT | Far End Crosstalk. |
| FFM | Flat Fade Margin. (Used in microwave system design.) |
| FFSK | Fast Frequency Shift Keying. |
| FHSS | Frequency Hopping Spread Spectrum. |
| FIPS | Federal Information Processing Standards. (Developed in US by NIST.) |
| FM | Frequency Modulation. (Analogue signal modualtion technique.) |
| FMFB | Frequency Modulation Feedback. |
| FPLMTS | Future Public Land Mobile Telecommunication System. (ITU-T name for third generation land mobile system, now renamed IMT2000. See also UMTS.) |

| | |
|---|---|
| FPODA | Fixed Priority Oriented Demand Assignment. (Medium multiple access method.) |
| FPS | Fast Packet Switch. (Standard for transmission based on frame relay or cell relay.) |
| FRA | Fixed Radio Access. |
| FSK | Frequency Shift Keying. (Digital modulation technique.) |
| FSN | Frequency Subset Number. (Number allocated to individual pagers within ERMES.) |
| FSS | Fixed Satellite Service. |
| FT | France Telecom. (French PTT.) |
| | |
| GAP | Groupe d'Analyse et de Prevision. (Analysis and Forecasting Group. A sub-committee of SOGT, part of the European Community.) |
| GATT | General Agreement on Tariffs and Trade. |
| GDN | Government Data Network. (UK private data network for use by government departments.) |
| GDP | Gross Domestic Product. (Measure of output from a country.) |
| GEN | Global European Network. (Joint venture between European PTOs to provide high speed leased line and switched services. Likely to be replaced by METRAN in mid-1990s.) |
| GEO | Geostationary Earth Orbit. (For satellites. Approximately 36000km altitude.) |
| GEOS | Geodetic Earth Orbiting Satellite. |
| GHz | GigaHerts. (Measure of frequency. Equal to 1000000000 cycles per second. See Hertz or Hz.) |
| GMSK | Gaussian Minimum Shift Keying. (Modulation technique, as used in GSM.) |
| GNP | Gross National Product. |
| GOES | Geostationary Operational Environmental Satellite. |
| GoS | Grade of Service. (Measure of service performance as perceived by the user.) |
| GOSIP | Government OSI Profile. (Government procurement standard.) |

| | |
|---|---|
| GPRS | General Packet Relay Service. (Data packet service in GSM.) |
| GPS | Global Positioning System. (Usually refers to satellite based vehicle positioning.) |
| GSC | Group Switching Centre. (Part of the hierarchy of switching in BT's network. Also called the primary trunk switching centre by ITU-T.) |
| GSLB | Groupe Special Large Bande. (A CEPT broadband working group.) |
| GSM | Global System for Mobile communication. (Previously Groupe Special Mobile. Pan-European standard for mobile communication.) |
| GSO | Geostationary Satellite Orbit. |
| GVPN | Global Virtual Private Network. |
| | |
| HACBSS | Homestead & Community Broadcasting Satellite Services. |
| HCI | Human Computer Interface. |
| HD | Harmonisation Document. (Sometimes used to describe an EN.) |
| HDB3 | High Density Bipolar 3. (Line transmission encoding technique.) |
| HDLC | Higher level Data Link Control. (ITU-T bit oriented protocol for handling data.) |
| HDSL | High bit rate Digital Subscriber Line. (Bellcore technical advisory for the transmission of high bit rate data over twisted copper lines.) |
| HDTV | High Definition Television. (New television transmission standard.) |
| HF | High Frequency. (Radio signal.) |
| HFC | Hybrid Fibre/Coax. (Distribution network, for example as used in cable TV.) |
| HKTA | Hong Kong Telecommunications Authority. |
| HLR | Home Location Register. (Database containing subscriber information in mobile communications.) |
| HOMUX | Higher Order Multiplexer. |

| | |
|---|---|
| HRC | Harmonically Related Carrier. (Carrier system used in cable television.) |
| HRDS | Hypothetical Reference Digital Section. (ITU-T G.921 for digital circuit measurements.) |
| HRP | Horizontal Radiation Pattern. (Of an antenna.) |
| HSE | Health and Safety Executive. (UK.) |
| Hz | Hertz. (Measure of frequency. One Hertz is equal to a frequency of one cycle per second.) |
| IAB | Internet Activities Board. |
| IBC | Integrated Broadband Communications. (Part of the RACE programme.) |
| IBCN | Integrated Broadband Communications Network. |
| IBM | International Business Machines. |
| IBS | Intelsat Business Services. |
| ICA | International Communications Association. (Telecommunications users' group in the USA.) |
| ICAO | International Civil Aviation Organisation. |
| ICMP | Internet Control Message Protocol. (Protocol developed by DARPA as part of Internet for the host to communicate with gateways.) |
| ICUG | International Closed User Group services. |
| IDN | Integrated Digital Network. (Usually refers to the digital public network which uses digital transmission and switching.) |
| IEC | International Electrotechnical Commission. |
| IEC | Interexchange Carrier. (US term for any telephone operator licensed to carry traffic between LATAs interstate or intrastate.) |
| IEEE | Institute for Electrical and Electronics Engineers. (USA professional organisation.) |
| IETF | Internet Engineering Task Force. |
| I-ETS | Interim European Telecommunicaitons Standard. (ETSI.) |
| IF | Intermediate Frequency. |
| IFIPS | International Federation of Information Processing Societies. |

| | |
|---|---|
| IFL | International Frequency List. (List of frequency allocations published by the ITU.) |
| IFRB | International Frequency Registration Board. (Part of the ITU's Radiocommunication Sector.) |
| IM | Intermodulation. |
| IMO | International Maritime Organisation. |
| IMSI | International Mobile Subscriber Identity. (Personal number associated with a PCN user's handset, and issued on his SIM. It is the number which the network uses to identify the mobile.) |
| IMTS | Improved Mobile Telephone System. (Analogue cellular radio standard, superseded by AMPS. North American origin.) |
| IN | Intelligent Network. |
| INMARSAT | International Maritime Satellite organisation. |
| INTELSAT | International Telecommunication Satellite organisation. |
| INTUG | International Telecommunications Users' Group. |
| IO | International Organisation. |
| IPM | Inter-Personal Messaging. (Use of Electronic Data Interchange.) |
| IPVC | International Private Virtual Circuit. |
| IR | Infrared. |
| IRC | Incrementally Related Carrier. (Carrier system used in cable television.) |
| ISD | International Subscriber Dialling. |
| ISDN | Integrated Services Digital Network. (Technique for the simultaneous transmission of a range of services, such as voice, data and video, over telephone lines.) |
| ISI | Inter-Symbol Interference. (Interference between adjacent pulses of a transmitted code.) |
| ISM | Industrial, Scientific and Medical. (Usually refers to ISM equipment or applications.) |
| ISO | International Standardisation Organisation. |
| ISO | International Satellite Organisation. (Includes Intelsat, Inmarsta and Eutelsat.) |

| | |
|---|---|
| ISO-DE | ISO Development Environment. |
| IT | Information Technology. (Generally refers to industries using computers e.g. data processing.) |
| ITAC | Information Technology Association of Canada. |
| ITAEGT | Information Technology Advisory and co-ordination Expert Group on private Telecommunication network standards. |
| ITANZ | Information Technology Association of New Zealand. |
| ITC | International Trade Commission. |
| ITC | Independent Television Commission. (UK regulator.) |
| ITSTC | Information Technology Steering Committee. (Comite de Direction de la Technologie de l'Information. Consists of CEN/CENLEC/ETSI.) |
| ITU | International Telecommunication Union. |
| ITU-R | International Telecommunication Union Radiocommunication sector. |
| ITU-T | International Telecommunication Union Telecommunication sector. |
| ITUSA | Information Technology User Standards Association. |
| IVA | Integrated Voice Application. |
| IVDS | Interactive Video and Data Services. |
| IVDT | Integrated Voice and Data Terminal. (Equipment with integrated computing and voice capabilities. In its simplest form it consists of a PC with telephone incorporated. Facilities such as storage and recall of telephone numbers is included.) |
| IVM | Integrated Voice Mail. |
| IVR | Interactive Voice Response. |
| IXC | Interexchange Carrier. (USA long distance telecommunication carrier.) |
| | |
| JEIDA | Japan Electronic Industry Development Association. |

| | |
|---|---|
| JESI | Joint European Standards Institute. (CEN and CENELEC combined group in the area of information technology.) |
| JIPDEC | Japan Information Processing Development Centre. |
| JIS | Japanese Industrial Standard. (Product marking used in Japan to denote conformance to a specified standard.) |
| JISC | Japanese Industrial Standards Committee. (Standards making body which is funded by the Japanese government.) |
| JSA | Japanese Standards Association. |
| JTM | Job Transfer and Manipulation. (Communication protocols used to perform tasks in a network of interconnected open systems.) |
| | |
| KBS | Knowledge Based System. |
| KDD | Kokusai Denshin Denwa Co. Ltd. (Japanese international carrier.) |
| kHz | KiloHertz. (Measure of frequency. Equals to 1000 cycles per second.) |
| | |
| LAN | Local Area Network. (A network shared by communicating devices, usually on a relatively small geographical area. Many techniques are used to allow each device to obtain use of the network.) |
| LASER | Light Amplification by Stimulated Emission of Radiation. (Laser is also used to refer to a component.) |
| LATA | Local Access and Transport Area. (Area of responsibility of local carrier in USA. When telephone circuits have their start and finish points within a LATA they are the sole responsibility of the local telephone company concerned. When they cross a LATA's boundary, i.e. go inter-LATA, they are the responsibility of an interexchange carrier or IEC.) |
| LBA | Least Busy Alternative. (Traffic routing strategy defined for fully connected networks.) |

| | |
|---|---|
| LBRV | Low Bit Rate Voice. (Speech encoding technique which allows voice transmission at under 64kbit/s.) |
| LCN | Local Communications Network. (ITU-T) |
| LCR | Least Cost Routing. (Usually applies to the use of alternative long distance routes, e.g. by using different carriers, in order to minimise transmission costs.) |
| LDM | Linear Delta Modulation. (Delta modulation technique in which a series of linear segments of constant slope provides the input time function.) |
| LEC | Local Exchange Carrier. (USA local telecommunication carrier.) |
| LEO | Low Earth Orbiting. (For satellites. Approximately 700km to 1500km altitude.) |
| LEOS | Low Earth Orbit System. (Satellite communication system with satellites not in geostationary orbit.) |
| LMSS | Land Mobile Satellite Services. |
| LOS | Line Of Sight. (Transmission system, e.g. microwave.) |
| LPC | Linear Predictive Coding. (Encoding technique used in pulse code modulation.) |
| LPTV | Low Power Television. |
| LRC | Longitudinal Redundanc Check. (Error checking procedure for transmitted data.) |
| LSB | Least Significant Bit. (Referring to bits in a data word.) |
| LTE | Line Terminating Equipment. (Also called Line Terminal Equipment. Equipment which terminates a transmission line.) |
| | |
| MA | Multiple Access. |
| MAC | Media Access Control. (IEEE standard 802. for access to LANs.) |
| MAC | Multiplexed Analogue Components. (Television transmission system.) |
| MAN | Metropolitan Area Network. |

| | |
|---|---|
| MASER | Microwave Amplification by Simulated Emission of Radiation. |
| MATV | Mast Antenna Television. (Or Master Antenna Television. Local cable television system for a hotel or apartment block.) |
| MBS | Mobile Broadband System. (RACE initiative.) |
| MCC | Mission Control Centre. |
| MCN | Micro-Cellular Networks. (Small cells used within cellular radio networks to provide PCN-type services.) |
| MCS | Maritime Communication System. |
| MD | Mediation Device. (ITU-T terminology for a device which carries out a protocol conversion function.) |
| MEO | Medium Earth Orbit. (For satellite. Approximately 10000km to 15000km altitude.) |
| MEP | Member of the European Parliament. |
| METEOSAT | Meteorological Satellite. |
| METRAN | Managed European Transmission Network. (CEPT initiative to provide a broadband backbone across Europe.) |
| MF | Multi Frequency. (A signalling system used with push-button telephones.) |
| MF | Mediation Function. (ITU-T term for a function involving protocol conversion.) |
| MFJ | Modification of Final Judgement. (Delivered by Judge Harold Greene. 1982 Act in the AT&T divestiture case.) |
| MHS | Message Handling System. (International standard.) |
| MHz | MegaHertz. (Measure of frequency. Equal to one million cycles per second.) |
| MIFR | Master International Frequency Register. (Register of allocated international frequencies maintained by the IFRB.) |
| MIPS | Million Instructions Per Second. (Measure of a computer's processing speed.) |

| | |
|---|---|
| MITI | Ministry of International Trade and Industry. (Japanese.) |
| MMI | Man Machine Interface. (Another name for the human-computer interface or HCI.) |
| MMIC | Monolithic Microwave Integrated Circuit. |
| MoD | Ministry of Defence. (UK.) |
| MODEM | Modulator/Demodulator. Device for enabling digital data to be send over analogue lines. |
| MoU | Memorandum of Understanding. |
| MPEG | Moving Picture Experts Group. |
| MPG | Microwave Pulse Generator. (A device for generating electrical pulses at microwave frequencies.) |
| MPT | Ministry of Posts and Telecommunications. (Japan.) |
| MSC | Main Switching Centre. (Part of the switching hierarchy in BT's network.) |
| MSC | Mobile Switching Centre. (Switching centre used in mobile radio systems.) |
| MSK | Minimum Shift Keying. (A form of frequency shift keying, or FSK.) |
| MSS | Mobile Satellite Service. |
| MTA | Message Transfer Agent. (The item which relays, stores and delivers messages within a Message Handling System, or MHS.) |
| MTBF | Mean Time Between Failure. (Measure of equipment reliability. Time for which an equipment is likely to operate before failure.) |
| MTN | Managed Transmission Network. |
| MTTR | Mean Time To Repair. (A measure of equipment availability. It is the time between an equipment failure and when it is operational again.) |
| MTX | Mobile Telephone Exchange. (Commonly used to describe a large exchange used within a cellular mobile system and connected to the PSTN.) |
| MUF | Maximum Usable Frequency. |
| | |
| NAFTA | North American Free Trade Association. |

| | |
|---|---|
| NAK | Negative Acknowledgement. (In data transmission this is the message sent by the receiver to the sender to indicate that the previous message contained an error, and requesting a re-send.) |
| NANP | North American Numbering Plan. (Telephone numbering scheme administered by Bellcore.) |
| NARS | North Atlantic Radio System. |
| NASA | National Aeronautics & Space Administration. (USA agency.) |
| NATA | North American Telecommunications Administration. |
| NBC | National Broadcasting Company. (USA.) |
| NBS | National Bureau of Standards. (USA.) |
| NCUF | National Computer Users Forum. |
| NDC | National Destination Code. (Part of numbering system.) |
| NE | Network Element. |
| NEP | Noise Equivalent Power. |
| NET | Nome Europeenne de Telecommunication. (European Telecommunications Standard, which is mandatory.) |
| NEXT | Near End crosstalk. The unwanted transfer of signal energy from one link to another, often closely located, at the end of the cable where the transmitted is located. |
| NF | Noise Figure. |
| N-ISDN | Narrowband Integrated Services Digital Network. |
| NIST | National Institute for Standards and Technology. (USA.) |
| NMC | Network Management Centre. |
| NMI | Network Management Interface. (Term used within OSI to indicate the interface between the network management system and the network it manages.) |
| NMT | Nordic Mobile Telephone. (Cellular system designed by the Nordic PTTs. NMT450 operates in the 450MHz band and NMT900 in the 900MHz band.) |

| | |
|---|---|
| NNI | Network Node Interface. (Usually the internal interfaces within a network. See UNI.) |
| NOAA | National Oceanic Atmospheric Administration. |
| NOI | Notice Of Inquiry. (FCC paper for comment.) |
| NPA | Numbering Plan Area. |
| NPP | Network Performance Parameters. |
| NPR | Noise Power Ratio. |
| NRA | National Regulatory Authority. |
| NRZ | Non Return to Zero. (A binary encoding technique for transmission of data.) |
| NRZI | Non Return to Zero Inverted. (A binary encoding techniqe for transmission of data.) |
| NSAP | Network Service Access Point. (Prime address point used within OSI.) |
| NT | Network Termination. (Termination designed within ISDN e.g. NT1 and NT2.) |
| NTE | Network Terminating Equipment. (Usually refers to the customer termination for an ISDN line.) |
| NTIA | National Telecommunications Industry Administration. (USA.) |
| NTN | Network Terminal Number. (Part of an international telephone number.) |
| NTSC | National Television System Committee. (USA.) |
| NTT | Nippon Telegraph and Telephone. (Japanese carrier.) |
| NTU | Network Terminating Unit. (Used to terminate subscriber leased line.) |
| NVOD | Near Video On Demand (television). |
| | |
| OA&M | Operations, Administration and Maintenance. (Also written as OAM.) |
| OAM&P | Operations, Administration, Maintenance & Provisioning. |
| OATS | Open Area Test Site. (Used for EMC measurements.) |
| OECD | Organisation for Economic Co-operation and Development. |

| | |
|---|---|
| OEM | Original Equipment Manufacturer. Supplier who makes equipment for sale by a third party. The equipment is usually disguised by the third party with his own labels.) |
| OFDM | Orthogonal Frequency Division Multiplexing. |
| OFTEL | Office of Telecommunications. (UK regulatory body.) |
| OKQPSK | Offset Keyed Quaternary Phase Shift Keying. |
| ONI | Operator Number Identification. (Operator used in a CAMA office to verbally obtain the calling number for calls originating in offices not equipped with ANI.) |
| OOK | On-Off Keying. (Digital modulation technique. Also known as ASK or Amplitude Shift Keying.) |
| OS | Operating System. (Or Operations System. ITU-T.) |
| OS | Outstation. |
| OSB | One Stop Billing. |
| OSF | Open Software Foundation |
| OSF | Operations System Function. (ITU-T.) |
| OSI | Open Systems Interconnection. (Refers to the seven layer reference model.) |
| OSIE | OSI Environment. |
| OSITOP | Open Systems Interconnection Technical and Office Protocol. |
| OSP | Operator Service Provider. (Company in the USA providing competitive toll operator services for billing and call completion.) |
| OSS | One Stop Shopping. |
| OTS | Orbital Test Satellite. |
| OWF | Optimum Working Frequency. |
| PABX | Private Automatic Branch Exchange. (PBX in which automatic connection is made between extensions.) |
| PAD | Packet Assembler/Disassembler. (Protocol converter used to provide access into the packet switched network.) |

| | |
|---|---|
| PAL | Pulse Alteration by Line. (Television encoding system.) |
| PAM | Pulse Amplitude Modulation. (An analogue modulation technique.) |
| PAMR | Public Access Mobile Radio. |
| PANS | Peculiar And Novel Services. (Often used in conjunction with POTS.) |
| PBX | Private Branch Exchange. (This is often used synonymously with PABX.) |
| PC | Personal Computer. |
| PC | Private Circuit. |
| PCM | Pulse Code Modulation. (Transmission technique for digital signals.) |
| PCMCIA | Personal Computer Memory Card International Association. |
| PCN | Personal Communications Networks. |
| PCS | Personal Communications Service. (North American term for the service provided over a PCN.) |
| PDA | Personal Digital Assistant. |
| PDH | Plesiochronous Digital Hierarchy. (Plesiochronous transmission standard.) |
| PDM | Pulse Duration Modulation. (Signal modulation technique, also known as Pulse Width Modulation or PWM.) |
| PDU | Protocol Data Unit. (Data and control information passed between layers in the OSI Seven Layer model.) |
| PEP | Peak Envelope Power. |
| PFD | Power Flux Density. (Measure of spectral emission strength.) |
| PFM | Pulse Frequency Modulation. (An analogue modulation technique.) |
| PHS | Personal Handy phone System. (Japanese cordless telephony standard.) |
| PIC | Personal Intelligent Communicator. |
| PIN | Personal Identification Number. (Security number used for items such as remote database entry.) |

## Acronyms 257

| | |
|---|---|
| PIN | Positive Intrinsic Negative. (Semiconductor structure used for some types of diodes.) |
| PLMN | Public Land Mobile Network. |
| PLP | Packet Level Protocol. |
| PM | Phase Modulation. (Analogue signal modulation technique.) |
| PMBX | Private Manual Branch Exchange. (PBX with connections between extensions done by an operator.) |
| PMR | Public Mobile Radio. |
| POCSAG | Post Office Code Standardisation Advisory Group. (Name given by study group to U.K. digital paging system. Adopted by ITU-T as Radio Paging Code RPC No. 1.) |
| PODA | Priority-Oriented Demand Assignment protocol. (Multiple access technique. See also FPODA and CPODA.) |
| POH | Path Overhead. (Information used in SDH transmission structures.) |
| PON | Passive Optical Network. (Technology for implementing fibre optic cable access in the local loop.) |
| POP | Point Of Presence. (Local access point into network, such as for the Internet. Also refers to the point of change over of responsibility from the local telephone company, within a LATA, to the long distance or inter-LATA carrier.) |
| POPSAT | Precise Orbit Positioning Satellite. |
| POR | Pacific Ocean Region. |
| POSI | Promoting conference for OSI. (Japan and Far East users' group active in functional standards and interconnection testing.) |
| POSP | Private Off-Site Paging. |
| POTS | Plain Old Telephone Service. (A term loosely applied to an ordinary voice telephone service.) |
| PPL | Phase Locked Loop. (Component used in frequency stability systems such as demodulators for frequency modulation.) |
| PPV | Pay Per View (television). |

| | |
|---|---|
| PPM | Pulse Phase Modulation. (An analogue modulation technique. Sometimes called Pulse Position Modulation.) |
| PRA | Primary Rate Access. (ISDN, 30B+D or 23B+D code.) |
| PRBS | Pseudo Random Binary Sequence. (Signal used for telecommunication system testing.) |
| PRF | Pulse Repetition Frequency. (Of a pulse train.) |
| PRK | Phase Reversal Keying. (A modification to the PSK modulation technique.) |
| PSDN | Packet Switched Data Network. (Or Public Switched Data Network. X.25 network, which may be private or public.) |
| PSE | Packet Switching Exchange. |
| PSK | Phase Shift Keying. (Analogue phase modulation technique.) |
| PSN | Packet Switched Network. |
| PSN | Public Switched Network. |
| PSPDN | Packet Switched Public Data Network. |
| PSS | Packet Switched Service. (Data service offered by BT.) |
| PSTN | Public Switched Telephone Network. (Term used to describe the public dial up voice telephone network operated by a PTT.) |
| PTC | Pacific Telecommunications Council. |
| PTN | Public Telecommunications Network. |
| PTO | Public Telecommunication Operator. (A licensed telecommunication operator. Usually used to refer to a PTT.) |
| PTT | Postal, Telegraph and Telephone. (Usually refers to the telephone authority within a country, often a publicly owned body. The term is also loosely used to describe any large telecomunications carrier.) |
| PUC | Public Utility Commission. (USA.) |
| PVC | Permanent Virtual Circuit. (Method for establishing a virtual circuit link between two nominated points. See also SVC.) |

| | |
|---|---|
| PWM | Pulse Width Modulation. (Analogue modulation technique in which the width of pulses is varied. Also called Pulse Duration Modulation, PDM, or Pulse Length Modulation, PLM.) |
| QA | Q interface Adaptor. |
| QAF | Q Adaptor Function. |
| QAM | Quadrature Amplitude Modulation. (A modulation technique which varies the amplitude of the signal. Used in dial up modems. Also known as Quadrature Sideband Amplitude Modulation or QSAM.) |
| QD | Quantising Distortion. |
| QoS | Quality of Service. (Measure of service performance as perceived by the user.) |
| QPRS | Quadrature Partial Response System. (Signal modulation technique.) |
| QPSK | Quadrature Phase Shift Keying. (Signal modulation technique.) |
| RA | Radiocommunications Agency. (UK body responsible for frequency allocation.) |
| RACE | Research and development in Advanced Communication technologies in Europe. |
| R&D | Research and Development. |
| RADARSAT | Radar Satellite. |
| RARC | Regional Administrative Radio Conference. |
| RBER | Residual Bit Error Ratio. (Measure of transmission quality. ITU-T Rec. 594-1.) |
| RBL | Re-Broadcast Link. |
| RBOC | Regional Bell Operating Company. (US local carriers formed after the divestiture of AT&T.) |
| RCU | Remote Concentrator Unit. |
| RF | Radio Frequency. (Signal.) |
| RFC | Request For Comment. |
| RFI | Radio Frequency Interference. |
| RIN | Relative Intensity Noise. (Measure of noise in an optical source.) |

| | |
|---|---|
| RITL | Radio In The Loop. (Radio used for the subscriber's local loop.) |
| ROES | Receive Only Earth Station. (For use with satellites.) |
| ROW | Right Of Way. (Usually refers to costs associated with laying cables.) |
| RPE-LTP | Regular Pulse Excitation Long Term Prediction. (Variant of LPC chosen as the speech encoding technique for GSM.) |
| RPFD | Received Power Flux Density. |
| RPOA | Recognised Private Operating Agencies. |
| RTNR | Real Time Network Routing. (Dynamic traffic routing strategy being implemented by AT&T.) |
| RTS | Request To Send. (Handshaking routine used in anlogue transmission, such as by modems.) |
| RTSE | Reliable Transfer Service Element. |
| RTT | Regie des Telegraphes et des Telephones. (Belgian PTT.) |
| RZ | Return to Zero. (A digital transmission system in which the binary pulse always returns to zero after each bit.) |
| | |
| SA | Standards Australia. (Australian national standards organisation.) |
| SAB | Services Ancillary to Broadcasting. |
| SAC | Special Access Code. (Special telephone numbers e.g. 800 service.) |
| SANZ | Standards Association of New Zealand. |
| SAR | Search And Rescue. |
| SARSAT | Search And Rescue Satellite. |
| SBS | Satellite Buisness Systems. (USA.) |
| SCADA | Supervisory Control and Data Acquisition. (Interface for network monitoring.) |
| SCC | Standards Council of Canada. |
| SCPC | Single Channel Per Carrier. (A transmission technique used in thin route satellite communication systems. See also SPADE.) |

## Acronyms 261

| | |
|---|---|
| SCVF | Single Channel Voice Frequency. (One of the signalling systems used in telex, i.e. ITU-T R.20.) |
| SDH | Synchronous Digital Hierarchy. |
| SDL | Specification Description Language. (ITU-T recommended language for specification and description of telecommunication systems.) |
| SDR | Speaker Dependent Recognition. (Speech recognition technique which requires training to individual caller's voice.) |
| SDU | Service Data Unit. (Data passed between layers in the OSI Seven Layer model.) |
| SDXC | Synchronous Digital Crossconnect. |
| SEA | Single European Act. (Amendment to the Treaty of Rome.) |
| SECAM | Sequence Coleur a Memoire. (Sequential Colour with Memory. French television encoding system.) |
| SED | Single Error Detecting code. (Transmission code used for detecting errors by use of single parity checks.) |
| SELV | Safety Extra Low Voltage circuit. (A circuit which is protected from hazardous voltages.) |
| SES | Severely Errored Second. (Any second with a BER exceeding $10^{-3}$, as per ITU-T G.821.) |
| SES | Ship Earth Station. |
| SESR | Severely Errored Second Ratio. |
| SHF | Super High Frequency. |
| SIM | Subscriber Identity Module. (Usually a plug in card used with a mobile radio handset.) |
| SIO | Scientific and Industrial Organisation. |
| SIP | Societa Italiana per l'Esercizio delle Telecomunicazion. (Italian PTT.) |
| SIR | Speaker Independent Recognition. (Speech recognition technique which does not need to be trained to individual caller's voice.) |
| SITA | Societe Internationale de Telecommunications Aeronautiques. (Refers to the organisation and its telecommunication network which is used by many |

of the world's airlines and their agents, mainly for flight bookings.)

| | |
|---|---|
| SLC | Subscriber Line Charge. (USA term for flat charge paid by the end user for line connection.) |
| SLIC | Subscriber Line Interface Card. (Circuitry which provides the interface to the network, usually from a central office switch, for digital voice transmission.) |
| SMATV | Satellite Master Television. (A MATV system relaying satellite TV.) |
| SMDR | Station Message Detail Recording. (Feature in a PABX for call analysis.) |
| SMDS | Switched Multimegabit Data Service. (High speed packet based standard proposed by Bellcore.) |
| SMR | Specialised Mobile Radio. (USA version of PMR.) |
| SMS | Short Messaging Service. (GSM feature which allocates channels for voice or data only when needed.) |
| SMS | Systeme Multiservice par Satellite. (Multi-service satellite system. A category of service offered by EUTELSAT.) |
| SMSC | Short Message Service Centre. (In GSM.) |
| SMSCB | Short Message Service Cell Broadcast. (Used for broadcasting information in GSM.) |
| SMX | Synchronous Multiplexer. |
| SNAP | Subnetwork Access Protocol. (IEEE protocol which allows non-OSI protocols to be carried within OSI protocols.) |
| SNI | Subscriber Network Interface. |
| SNMP | Simple Network Management Protocol. (Network management system within TCP/IP.) |
| SNR | Signal to Noise Ratio. |
| SNV | Association Suisse de Normalisation. (Swiss standards making body.) |
| SOGITS | Senior Officals Group on Information Technology Standards. (Part of the European Community.) |
| SOGT | Senior Officials Group on Telecommunications. (Part of European Community, composed of Minis- |

| | |
|---|---|
| | ters of Telecommunications and Industry from member states.) |
| SOHO | Small Office Home Office (market). |
| SONET | Synchronous Optical Network. (Synchronous optical transmission system first developed in North America, and developed by ITU-T into SDH.) |
| SPADE | Single channel per carrier Pulse code modulation multiple Access Demand assignment Equipment. (A SCPC technique.) |
| SPC | Stored Program Controller. (Usually refers to a digital exchange.) |
| SPEC | Speech Predictive Encoding Communications. |
| SQNR | Signal Quantisation Noise Ratio. |
| SSB | Single Sideband. |
| SSBSC | Single Sideband Suppressed Carrier modulation. (A method for amplitude modulation of a signal.) |
| SS-CDMA | Spread Spectrum Code Division Multiple Access. |
| SSMA | Spread Spectrum Multiple Access. |
| SSSO | Specialised Satellite Service Operator. |
| SSTDMA | Satellite Switched Time Division Multiple Access. |
| SSTV | Slow Scan Television. |
| STAR | Special Telecommunications Action for Regional Development. (European programme.) |
| STD | Subscriber Trunk Dialling. |
| STDM | Synchronous Time Division Multiplexing. |
| STDMX | Statistical Time Division Multiplexing. |
| STE | Signalling Terminal Equipment. |
| STM | Synchronous Transport Module. (Basic carrier module used within SDH, e.g. STM-1, STM-4 and STM-16.) |
| STMR | Sidetone Masking Rating. (Measure of talker effects of sidetone.) |
| STX | Synchronous Transmission crossconnect. (Cross-connect used within SHD.) |
| STX | Start of Text. (Control character used to indicate the start of data transmission. It is completed by a End of Text character, or ETX.) |

| | |
|---|---|
| SVC | Switched Virtual Circuit. (Method for establishing any to any virtual circuit link. See also PVC.) |
| TA | Terminal Adaptor. (Used within ISDN to convert between non-ISDN and ISDN references.) |
| TA | Telecommunication Authority. |
| TA 84 | Telecommunications Act of 1984. (UK.) |
| TACS | Total Access Communication Systems. (Adaption of AMPS by the UK to suit European frequency allocations.) |
| TAPI | Telephony Applications Programming Interface. (A standard for linking telephones to PCs.) |
| TASI | Time Assignment Speech Interpolation. (Method used in analogue speech transmission where the channel is activated only when speech is present. This allows several users to share a common channel.) |
| TC | Transport Class. (E.g. TC 0, TC 4, etc.) |
| TCD | Technical Co-operation Department. |
| TCM | Time Compression Multiplexing. (Technique which separates the two directions of transmission in time.) |
| TCM/DPSK | Trellis Coded Modulation/Differential Phase Shift Keying. |
| TCP/IP | Transmission Control Protocol/Internet Protocol. (Widely used transmission protocol, originating from the US ARPA defence project.) |
| T-DAB | Terrestrial Digital Audio Broadcasting. |
| TDD | Time Division Duplex transmission. |
| TDD/FDMA | Time Division Duplex/Frequency Division Multiple Access. (Means of multiplexing several two way calls using many frequencies, with a single two way call per frequency.) |
| TDD/TDMA | Time Division Duplex/Time Division Multiple Access. (Multiplexing two way calls using a single frequency for each call and multiple time slots.) |

| | |
|---|---|
| TDM | Time Division Multiplexing. (Technique for combining, by interleaving, several channels of data onto a common channel. The equipment which does this is called a Time Division Multiplexer.) |
| TDMA | Time Division Multiple Access. (A multiplexing technique where users gain access to a common channel on a time allocation basis. Commonly used in satellite systems, where several Earth stations have total use of the transponder's power and bandwidth for a short period, and transmit in bursts of data.) |
| TDRS | Tracking and Data Relay Satellite. |
| TE | Terminal Equipment. |
| TEDIS | Trade Electronic Data Interchange System. (European Commission programme.) |
| TEI | Terminal Endpoint Identifier. (Used within ISDN Layer 2 frame.) |
| TELMEX | Telefonos de Mexico. (Mexican PTT.) |
| TELR | Talker Echo Loudness Rating. (Overall loudness rating of the talker echo path.) |
| TEM | Transverse Electromagnetic cell. (Code used for measuring characteristics of receivers such as pagers.) |
| TEMA | Telecommunication Equipment Manufacturers Association. (UK.) |
| TETRA | Trans European Trunked Radio. (ETSI standard for PMR.) |
| TFTS | Terrestrial Flight Telephone Service. (ETSI terminology for airborne communication systems.) |
| TIA | Telecommunication Industry Association. (US based. Formed from merger of telecommunication sector of the EIA and the USTSA.) |
| TINA | Telecommunication Information Network Architecture. |
| TM | Trade Mark. |
| TMA | Telecommunication Managers' Association. (UK) |
| TMN | Telecommunications Management Network. |

| | |
|---|---|
| TMR | Trunked Mobile Radio. |
| TNV | Telecommunication Network Voltage circuit. (Test circuit for definition of safety in telecommunication systems.) |
| TOA | Take Off Angle. (Refers to antenna systems.) |
| TOT | Telephone Organisation of Thiland. (Thai carrier.) |
| TPON | Telephony over Passive Optical Networks. |
| TR | Technical Report. (ISO technical document; not a standard.) |
| TRAC | Technical Recommendations Application Committee. (Comite Charge de l'Application des Recommendations Techniques. Part of CEPT.) |
| TSAPI | Telephony Server Applications Programming Interface. (A standard for linking servers to PABXs.) |
| TT&C | Tracking, Telemetry & Command. (Satellite antenna.) |
| TTC | Telecommunications Technology Committee (Japanese.) |
| TTE | Telecommunication Terminal Equipment. |
| TUA | Telecommunications Users' Association. (UK.) |
| TUF | Telecommunications Users' Foundation. (UK.) |
| TV | Television. |
| TVRO | Television Receive Only. (Domestic equipment for the reception of television via satellite.) |
| TeraFLOP | Trillion Floating Point Operations per second. (Measure of a super computer.) |
| | |
| UAP | User Application Process. |
| UART | Universal Asynchronous Receiver/Transmitter. (The device, usually an integrated cirucit, for transmission of asynchronous data. See also USRT and USART.) |
| UDF | Unshielded twisted pair Development Forum. (Association of suppliers promoting transmission over UTP.) |
| UDP | User Datagram Protocol. |

| | |
|---|---|
| UHF | Ultra High Frequency. (Radio frequency, extending from about 300MHz to 3GHz.) |
| UI | User Interface. |
| UL | Underwriters Laboratories. (Independent USA organisation involved in standards and certification.) |
| UMTS | Universal Mobile Telecommunications System. (ETSI terminology for a future public land mobile telecommunications system.) |
| UN | United Nations. |
| UNCTAD | United Nations Conference on Trade And Development. |
| UNI | User Network Interface. (Also called User Node Interface. External interface of a network.) |
| UPS | Uninterrupted Power Supply. (Used where loss of power, even for a short time, cannot be tolerated.) |
| UPT | Universal Personal Telecommunications. (ITU-T concept of the personal telephone number.) |
| USART | Universal Synchronous/Asynchronous Receiver/Transmitter. (A device, usually an integrated circuit, used in data communication devices, for conversion of data from parallel to serial form for transmission.) |
| USB | Upper Sideband. |
| USITA | US Independent Telephone Association. |
| USRT | Universal Synchronous Receiver/Transmitter. (A device, usually an integrated circuit, which converts data for transmission over a synchronous channel.) |
| USTA | US Telephone Association. |
| USTSA | US Telecommunication Suppliers Association. |
| UTP | Unshielded Twisted Pair. (Cable.) |
| VAD | Voice Activity Detection. (Technique used in transmission systems to improve bandwidth utilisation.) |
| VADIS | Video Audio Digital Interactive System. |
| VADS | Value Added Data Service. |
| VAN | Value Added Network |
| VANS | Value Added Network Services. |

| | |
|---|---|
| VAS | Value Added Service. (See also VANS.) |
| VASP | Value Added Service Provider. |
| VBR | Variable Bit Rate. |
| VDT | Video Dial Tone. (System in which subscriber can obtain variety of video services by dialling through the PSTN. Also referred to as VOD or Video on Demand.) |
| VF | Voice Frequency. (Signalling method using frequencies within speech band. Also called in-band signalling. Also refers to the voice frequency band from 300Hz to 3400Hz.) |
| VFCT | Voice Frequency Carrier Telegraph. |
| VHF | Very High Frequency. (Radio frequency in the range of about 30MHz and 300MHz.) |
| VLF | Very Low Frequency. (Radio frequency in the range of about 3kHz to 30kHz.) |
| VLR | Visitor Location Register. (Database associated with a MCS in a mobile radio system, containing information of its current users.) |
| VMS | Voice Messaging System. |
| VNL | Via Net Loss. (The method used to assign minimum loss in telephone lines in order to control echo and singing.) |
| VOA | Voice Of America. |
| VOD | Video On Demand. (Also known as VDT or Video Dial Tone.) |
| VPN | Virtual Private Network. (Part of a network operated by a public telephone operator, which is used as a private network.) |
| VQ | Vector Quantisation. (Encoding method.) |
| VQL | Variable Quantising Level. (Speech encoding method for transmission of speech at 32kbit/s.) |
| VRC | Vertical Redundancy Check. (Parity method used on transmitted data for error checking.) |
| VRP | Vertical Radiation Pattern. (Of an antenna.) |
| VSAT | Very Small Aperture Terminal. (Satellite receiver.) |

| | |
|---|---|
| VSB | Vestigial Sideband modulation. (A method for amplitude modulation of a signal.) |
| VSWR | Voltage Standing Wave Ratio. |
| VT | Virtual Tributary. (SONET terminology for a virtual container.) |
| VTP | Virtual Terminal Protocol. |
| | |
| WAN | Wide Area Network. |
| WARC | World Administrative Radio Conference. (Of ITU. Now known as WRC.) |
| WATTC | World Administrative Telephone and Telegraph Conference. (Of ITU.) |
| WBLLN | Wideband Leased Line Network. |
| WDM | Wavelength Division Multiplexing. (Multiplexing technique used with optical communications systems.) |
| WDMA | Wavelength Division Multiple Access. (Multiple access technique.) |
| WILL | Wireless In the Local Loop. (Local loop access method.) |
| WIMP | Windows, Icons, Mouse and Pointer. (Display and manipulation technique for graphical interfaces, e.g. as used for network management.) |
| WLAN | Wireless Local Area Network. |
| WMO | World Meteorological Organisation. |
| WRC | World Radio Conference. (Of ITU.) |
| WS | Work Station. |
| WSF | Workstation Function. |
| WTO | World Trade Organisation. |
| WWW | World Wide Web. (Internet.) |
| | |
| ZVEI | Zurerein der Electronissches Industrie. |
| ZZF | Zentrasamt fur Zulassungen im Fernmeldewessen. |

# Index

A3E (AM telephony), 65
Absence rack units, 16
Acronyms, 226-69
Address code word, 72
Advanced Mobile Phone System (AMPS), 121-2
Aloha messages, 60, 78
Analogue signalling, 65-70
  CT standards, 82-4
Antennas, design, 30
APOC code format, 20
Attitude stabilisation, satellite, 190-91
Authentication procedures, GSM system, 136-7

Band III system, 58-9
  call processing, 60-61
  control channel, 60
  data messaging, 62
  half/full duplex, 59
Base station
  communal, 51
  coverage, 56
  PCN design, 166-72
  remote, 51
Base Station Controller (BSC), 152-3, 155, 167-9
Base Station Subsystem (BSS), 167-9

Base Transceiver Station (BTS), 152, 155, 167-8
Batteries
  pagers, 33-4
  portable, 47
Belgium, frequency allocations, 5
Bi-phase level modulation 24, 25
British Approvals Board for Telecommunications (BABT), 12
Broadcast channels, 163-4
Broadcast Common Control Channel (BCCH), 161, 164
Broadcasting, satellite *see* Satellite broadcasting

C450 radio system, 124-5
Call logging facility, 16
Call set up, PCN, 153-4
Canada, frequency allocations, 6
Cell clusters, trunking systems, 55-6
Cell sectorisation, 118-19
Cellular Data Link Control (CDLC), 142
Cellular radio systems
  adding capacity, 117-19
  alternative systems, 119-20

## Index

AMPS, 121-2
C450, 124-5
call set up, 109-10
cell patterns, 112-14
cell splitting, 117-18
data services, 141-3
developments, 143-5
GSM *see* GSM cellular system
in-call handover, 111
interference control, 112-19
location registration, 109
networks, 107-108
NMT, 123-4
power control, 112
services available, 138-40
signalling, 108-109
TACS, 122-3
value added services, 140-41
Co-channel interference, 114
Code Division Multiple Access (CDMA), 145, 208
Code formats, on site paging systems, 7-8
Common air interface (CAI) standardisation, 88
Common Control Channel (CCCH), 164
Common Interface (CI) specification, 96
Communication satellite systems
for broadcasting, 218-20
communication chain, 193-5
developments, 220-21
history, 175-6
international regulations, 176-82
for mobile stations, 217-218
modulation techniques, 203-205
multiple access, 205-208
*see also* Spacecraft technology
Concatenated messages, 72, 73
Conference of European Posts and Telecommunications (CEPT)
analogue CT standard, 83-4
GSM standard, 125, 128
Continuous Tone Controlled Squelch System (CTCSS), 12, 66-7
Control Channel System Codeword (CCSC), 60
Control channels, 57, 163-4
Band III system, 60
GSM system, 129
Cordless business communication system (CBCS), 85
Cordless communications, 82
CT2/CAI standard (CT2 with Common Air Interface), 87-8
codec algorithm, 94
frame multiplex structure, 90-92
implementation, 94
protocol structure, 92-3
radio aspects, 88-90
transmission plan, 93

Data channels, 163
Data code word, 72, 73
Data services, cellular systems, 141-3

Data transmission, PMR, 70-77
DCS 1800 (Digital Cellular System at 1800MHz), 150-51
  handover types, 162
  standard, 145
Decoder and audio block diagram, 29
DECT (Digital European Cordless Telecommunications) standard, 87, 948
  codec standard, 105
  frame structure, 104
  interworking units (IWU), 100
  operational features, 103-104
  protocol, 98-100
  radio parameters, 101-103
  spectrum resource, 100-101
  transmission plan, 105
Dedicated Control Channel, 164
Denmark, frequency allocations, 5
Digital Cross Connect (DCX) equipment, 157
Digital modulation, 24, 25
Digital speech codec algorithm, CT2/CAI, 94
Digital technology, 84-5
  application, 85-6
  code formats, 20
  standards, 86-7
Direct conversion receiver, 28
Discontinuous reception, 171
Discontinuous transmission (DTX), 171
Diversity, space, 170

Dual Tone Medium Frequency (DTMF), 65-6
Dynamic channel selection (DCS), 84, 88
Dynamic power control, 171

Echo control/cancellation, 105, 166
EEPROM, as programmable pager memory, 34-5
Electronic Security Number (ESN), 61
Encoding, data transmission, 74
Encryption, 166
Equipment Identity Register (EIR), GSM system, 137
Erlang (measure of traffic), 53-4, 86
ERMES system 3842, 43
  code format, 20
Error checking, data transmission, 74
Euromessage system, 37-8, 39
European Community Directives, specification harmonisation, 26
European Selective Paging Association (ESPA), 12
European Telecommunications Standards, 86-7
Eurosignal system, 19
EUTELSAT system, 214
  EUTELSAT II satellite, 200, 214

F3E (narrowband telephony), 65
Facsimile transmission, 142, 143

Fading effects, 8, 66, 68, 70, 115-16
Fast Associated Control Channel (FACCH), 164
Fast Frequency Shift Keying (FFSK), 69-70
Federal Communications Commission (FCC), on site paging regulations, 12
Figure of merit (G/T), antenna, 204, 211, 212, 217
FLEX code format, 20
Forward Control Channel (FOCC), 57
Forward error correction (FEC), 204
Forward scatter *see* Troposcatter
Frame format, data transmission, 75-6
France, frequency allocations, 3-4
Free field sensitivity, 32
Frequency allocations
  constraints, 180-82
  paging systems, 26
  PCN, 169-72
  satellite links, 176-9
Frequency Correction Channel (FCCH), 163
Frequency division multiple access (FDMA), 205-206
Frequency division multiplex (FDM), 204
Frequency generation, PMR, 47
Frequency hopping, 170-71
Frequency Shift Keying (FSK), 24

Full duplex system, 49-51

Geostationary orbit (GSO), 194-5
Germany
  frequency allocations, 3
  private off site paging system, 36
  on site paging regulations, 12
Global title, 160
GSM cellular system, 125-6
  air interface, 128-31
  architecture, 126-7
  GSM 900 base station, 166
  roaming, 138
  security features, 136-7
  services, 137-8
  signalling, 133-4
  speech coding, 131-3
  standard for PCN, 150-51

Half duplex system, 49
Half/full duplex, Band III system, 59
Handoff usage, 57, 58
Handover procedures, PCN, 162, 172
Hata model formula (path loss), 116
High definition television (HDTV), 221
Holland
  frequency allocations, 4
  Semafoon network, 19
Home Location Register (HLR), 153

Information services, cellular systems, 140

INMARSAT system, 176, 217-18
Intelligent Network (IN) techniques, 144-5
INTELSAT system, 175, 211-15
  INTELSAT IV satellite, 197, 199
  INTELSAT V satellite, 190, 191, 202
  INTELSAT VI satellite, 200, 202
International mobile subscriber identity (IMSI), 159
Italy, frequency allocations, 4

J3E (single sideband telephony), 65

Manchester code *see* Bi-phase level modulation
MARISAT satellites, 176, 217
'Meet me' facility, 10
Messaging services, cellular systems, 140
Microcellular techniques, 143-4
  PCN, 172-3
Mobile roaming, 567
Mobile station ISDN number (MSISDN), 159
Mobile switching centre (MSC), 153, 167
Mobilsoekring system (MBS), 24-5

Modulation techniques, communication satellite systems, 203-205
MPT1317 code of practice, 71-3
MPT1327
  sample command structure, 78-9
  signalling specification, 58
Multipath equalisation, 171-2
Multiplexer structures, CT2/CAI, 90-92

Networks
PCN *see* Personal Communications Networks (PCN), network overview
trunk, 158-9
Netz-C system *see* C450 system
Non-transparent data services, 142-3
Nordic Mobile Telephone (NMT) systems, 123-4
NRZ level modulation, 24, 25
NRZ mark modulation, 24, 25

On site paging systems
  applications, 7
  code formats, 7-8
  large, 14-19
  regulations, 12
  small, 8-9
Open Systems Interconnection (OSI) model, 92, 98, 99
Overlay paging, 35

Packet data communication, 74-5

# Index

Paging receivers
  alphanumeric, 6
  antennas, 30
  architecture, 27-30
  design, 26
  numeric, 6
  performance measurement, 30-32
  programming, 34-5
  specifications, 26, 27
  tone type, 6
Paging systems
  frequency allocations, 2-6
  markets, 1-2
  zones, 23-4
*see also under* On site paging systems; Wide area paging systems
Personal Communications Networks (PCN), 145
  base station design, 166-72
  call set up, 153-4
  definition, 150-51
  history, 148-50
  mobility, 160-62
  network overview, 151-3
  numbering systems, 159-60
  PCN1800 base station, 166
  planning, 154-60
  radio channel coding, 163-6
  use of microcells, 172-3
Personal Communications Services (PCS), 145
Phase shift keying (PSK), 204
Point codes, 160

Post Office Code Standardisation Advisory Group (POCSAG) code format, 18, 20-23
Power sources, pagers, 33-4
PRE-LPC coder, 164
Private interconnect service, cellular systems, 140
Private Mobile Radio (PMR), 46
  alternative systems, 48-51
  Band III system *see* Band III system
  development, 78
  equipment, 46-8
  overlay paging, 35
  spectrum usage, 46, 53
  trunked, 53-8
  usage, 77
Private off site paging system (POSP), 36
Programmable pager memory, 34-5
Public Access Mobile Radio (PAMR), usage, 77

Quasi-synchronous operation, 51-3

Radio Paging Association, UK, 36
Radio Paging Code RPC No 1 *see* Post Office Code Standardisation Advisory Group code format
Radio planning, 116-17
  cell repeat patterns, 112-14
Radio Signal Strength Indicator (RSSI), 56

Radio subsystem (RSS), 160
Rayleigh fading *see* Fading effects
Regular excited linear predictive coder (PRE-LPC), 164
Regulations
  satellite links, 176-82
  on site paging systems, 12
Reverse Control Channel (RECC), 57
Revertive paging, 35-6
Ring configuration, BTS, 156
Ring-star configuration, BTS, 157
Roaming
  allocated number, 160
  GSM system, 138
  National, 150
RPE-LTP algorithm, 168

Satellite links *see* Communication satellite systems
Satellites
  antennas, 201-202
  applications, 208-10
  footprint, 202
  indirect paging, 42, 44
  land mobile data communication, 76-7
  *see also* Spacecraft technology
Secondary calling *see* Revertive paging
Secondary Station Identifier (SSID), 75
Security, GSM system, 136-7
Semafoon network, 19
Sequential tone signalling, 68-9
Short data messages (SDM2), 62
Signal processing
  GSM system, 133-4
  PMR, 47
Simplex system, 48
SkyTel service, 42
Slow Associated Control Channel (SACCH), 164
Space Shuttle, 184
Space-earth transmission loss, 195-7
Spacecraft technology
  attitude stabilisation, 190-91
  launching, 184-5
  orbital perturbations, 185, 189-90
  orbits, 182-3
  power supplies, 192
  telemetry, tracking and command (TT&C), 192-3
Spectrum resource, DECT system, 100-101
Speech channels, 163
Speech coding
  algorithm, 88
  GSM system, 131-3
  PCN, 164-6
Standalone Dedicated Control Channel (SDCCH), 164
Star configuration, BTS, 155-6
Status messages (SDM1), 62
Subscriber Identity Module (SIM), 159
Superhet block diagram, 28
Sweden
  frequency allocations, 5-6

# Index

Mobilsoekring system (MBS), 24-5
Synchronisation Channel (SCH), 163

Talk back paging, 12-13
Telemetry, tracking and command (TT&C) facilities, spacecraft, 192-3
Telephone coupling, 10-11
Telepoint service, 85
Thin route telecommunications, 215-216
Time Division Multiple Access (TDMA), 128, 163, 206-208
Time Division Multiplex (TDM), 204
Tone code formats, wide area paging systems, 19
Tones, international standards, 69
Total Access Communications System (TACS), 122-3
Traffic channels, 163
Transmission standards, PMR, 65
Transmitters
  on site, 16-18
  synchronisation, 18-19
Transparent data services, 142
Transponders, satellite, 197-201
Transverse ElectroMagnetic (TEM) cell, 31-2
Travelling wave tube (TWT), 205
Troposcatter, 62-3
  equipment, 63
  signal levels, 64
Tropospheric ducting, 64-5
Trunk telecommunications, by satellite, 209, 211-15
Trunking System Controller (TSC), 61
Trunking systems, 53-5
  area coverage, 55
Tuned circuits, PMR, 47

UK
  analogue CT standard, 83
  British Approvals Board for Telecommunications, 12
  frequency allocations, 2-3
  Radio Paging Association, 36
USA
  frequency allocations, 6
  on site paging regulations, 12

Value added services, cellular systems, 140
Visitor Location Register (VLR), 153
Voice Activity Detection (VAD), 171

Wide area paging systems
  digital code formats, 20
  private, 35-6
  tone code formats, 19
Wide area transmission, 23-4
World Administration Radio Conference, 46

ZZF, on site paging regulations, 12